HAMILTONIAN SUBMANIFOLDS OF REGULAR POLYTOPES

Von der Fakultät Mathematik der Universität Stuttgart

zur Erlangung der Würde eines Doktors der

Naturwissenschaften (Dr. rer. nat.) genehmigte Abhandlung

Vorgelegt von

Dipl.-Math. Felix Effenberger

geboren in Frankfurt am Main

Hauptberichter: Prof. Dr. Wolfgang Kühnel (Universität Stuttgart)

Mitberichter: apl. Prof. Dr. Wolfgang Kimmerle (Universität Stuttgart)

Mitberichter: Prof. Dr. Michael Joswig (Universität Darmstadt)

Mitberichter: Prof. Isabella Novik, PhD (University of Washington)

Tag der mündlichen Prüfung: 23. Juli 2010

Institut für Geometrie und Topologie der Universität Stuttgart

2010

Bibliografische Information der Deutschen Nationalbibliothek

Die Deutsche Nationalbibliothek verzeichnet diese Publikation in der
Deutschen Nationalbibliografie; detaillierte bibliografische Daten sind
im Internet über http://dnb.d-nb.de abrufbar.

D 93

ISBN 978-3-8325-2758-7

Logos Verlag Berlin GmbH
Comeniushof, Gubener Str. 47,
10243 Berlin
Tel.: +49 (0)30 42 85 10 90
Fax: +49 (0)30 42 85 10 92
INTERNET: http://www.logos-verlag.de

To my family

Acknowledgments

First of all, I want to thank *Prof. Wolfgang Kühnel* in manifold ways. It was his passion for mathematics and the field of combinatorial topology that made me choose to continue working in the field after being a research assistant for him during my time as a student. He was a caring and supporting supervisor, constantly encouraging me in my research. His door was always open and many times he had just the right hint that brought me back on track when I felt stuck with a problem. Looking back on my time as his PhD student, I am convinced that I could hardly have made a better choice regarding the subject and supervisor of my thesis.

Furthermore, I want to thank my dear friend and colleague *Dipl.-Math. Jonathan Spreer* for his company, his patience and the fun times we have had when working on common mathematical projects like `simpcomp` or helping each other out with difficult problems. Without him, my time here at the University of Stuttgart would have been a lot less fun.

In addition, I want to thank all other members of the Institute of Geometry and Topology at the University of Stuttgart, especially *apl. Prof. Wolfgang Kimmerle* for our enlightening discussions about group- and representation-theoretic topics.

I thank *Priv.-Doz. Frank H. Lutz* for his invitations to the Technical University of Berlin, his generous hospitality and for pointing out to me and inviting me to various conferences.

Moreover, I thank *Prof. Edward Swartz* for his hospitality and candidness during my stay at Cornell University – although my visit in Ithaca was not a long one, I immediately felt at home and enjoyed the mathematical discussions with him and the other members of his group.

I would also like to thank the reviewers of the work at hand, especially *Prof. Isabella Novik* for her helpful hints concerning the revision of Section 5.3.1.

Last but not least, I want to thank my family and friends for their constant patience and confidence that had a substantial influence on the success of this work.

This dissertation was supported and funded by the German Research Foundation (Deutsche Forschungsgemeinschaft), grant Ku-1203/5-2, the University of Stuttgart and the German National Academic Foundation (Studienstiftung des Deutschen Volkes). My trip to Cornell University was funded by the German Academic Exchange Service (Deutscher Akademischer Austauschdienst).

Contents

Notation and Symbols

$\varnothing$	empty set
$\#M$	cardinality of a set M
$\sqcup$	disjoint union
$\mathbb{N} = \{1, 2, 3, \ldots\}$	set of natural numbers
$\mathbb{Z}$	set of integers (also as ring)
$\mathbb{Z}_n$	factor ring of $\mathbb{Z}$ by the ideal $n\mathbb{Z}$ (also as ring)
$\overline{a}$	residue class of the integer a in $\mathbb{Z}_n$
$\mathbb{Q}, \mathbb{R}, \mathbb{C}$	sets of rational, real and complex numbers (also as fields)
$\mathbb{F}$	arbitrary field
$\mathbb{F}_q$	finite field on q elements
$\mathbb{F}^n$	n-dimensional vector space over the field $\mathbb{F}$
E^n	n-dimensional Euclidean space
Δ^d	d-simplex
$\langle v_1, \ldots, v_{d+1} \rangle$	d-simplex on vertex set $v_1, \ldots, v_{d+1}$ (also abstract)
β^d	d-cross polytope (or d-octahedron)
C^d	d-cube
P^d, P	convex $(d\text{-})$polytope P
∂P	boundary of a convex polytope P
$C(P)$	polytopal complex of a convex polytope P
$C(\partial P)$	boundary complex of a convex polytope P
$f(C)$	f-vector of a polytopal complex C
$h(C)$	h-vector of a simplicial complex C
$\chi(C)$	Euler characteristic of a polytopal complex C
∂C	boundary of a polytopal complex C

$\mathrm{Aut}(C)$	automorphism group of a polytopal C
$|C|$	underlying set of a polytopal complex
$V(C)$	vertex set of a polytopal complex
$\mathrm{skel}_k(C)$	k-skeleton of the polytopal complex C
$\mathrm{lk}_C(\sigma)$	link of the face σ in the polytopal complex C
$\mathrm{st}_C(\sigma)$	star of the face σ in the polytopal complex C
$\mathrm{span}_C(X)$	span of the vertex set $X \subseteq V(C)$ in the polytopal complex C
$C_1 \# C_2$	connected sum of two simplicial complexes C_1, C_2
$C^{\#k}$	k-fold connected sum of a simplicial complexes C with itself
$C_1 \times C_2$	cartesian product of two simplicial complexes C_1, C_2
$C_1 \mathbin{\rtimes} C_2$	twisted product of two simplicial complexes C_1, C_2
$\Phi_A(C)$	bistellar move on the simplicial complex C
$C_1 * C_2$	join of two vertex-disjoint (abstract) simplicial complexes C_1 and C_2
∂	boundary operator
$H_*(C; G)$	homology groups of the simplicial complex C with coefficients in G
$H_*(C)$	homology groups of the simplicial complex C with coefficients in $\mathbb{Z}$
$H_*(C, D; G)$	relative homology groups of the simplicial complex C with a subcomplex C with coefficients in G
$H_*(C, D)$	relative homology groups of the simplicial complex C with a subcomplex C with coefficients in $\mathbb{Z}$
$\beta_i(C; G)$	i-th Betti number of a simplicial complex C with respect to the group of coefficients G
$\beta_i(C)$	i-th Betti number of a simplicial complex C with respect to the group of coefficients $\mathbb{Z}$
$C_1 \cong C_2$	combinatorial equivalence of two polytopal complexes C_1, C_2

$\mu_i(f; \mathbb{F})$	number of critical points of index i of the rsl function f with respect to the field $\mathbb{F}$
$\mu_i(f)$	number of critical points of index i of the rsl function f with respect to the field $\mathbb{F}_2$
$G_1 \times G_2$	direct product of the groups G_1, G_2
$G_1 \rtimes G_2$	semi direct product of the groups G_1, G_2
$(a_1 \cdots a_n)$	cycle of length n
S_n	symmetric group of degree n
C_n	cyclic group of order n
$G_1 \cong G_2$	isomorphy of two groups G_1, G_2
$\binom{n}{k}$	binomial coefficient "n choose k"
$a\vert b$	for $a, b \in \mathbb{Z}$ and b is an integer multiple of a
$\{q_1, \ldots, q_{d-1}\}$	Schläfli symbol of a regular d-polytope
$\mathcal{K}(d)$	Walkup's class of manifolds with stacked vertex links
$\mathcal{K}^k(d)$	class of triangulated manifolds with k-stacked vertex links
$\delta B = (d_1, \ldots, d_n)$	difference sequence B of length n
$\partial D = (d_1 : \cdots : d_n)$	difference cycle D of length n
$\Sigma_l^{l+m}(\partial D)$	running sum of a difference cycle ∂D
$\searrow$	inheritance relation on difference cycles

Abstract

This work is set in the field of *combinatorial topology*, a mathematical field of research in the intersection of the fields of topology, geometry, polytope theory and combinatorics.

This work investigates polyhedral manifolds as subcomplexes of the boundary complex of a regular polytope. Such a subcomplex is called *k-Hamiltonian*, if it contains the full k-skeleton of the polytope. Since the case of the cube is well known and since the case of a simplex was also previously studied (these are so-called *super-neighborly triangulations*), the focus here is on the case of the cross polytope and the sporadic regular 4-polytopes. By the results presented, the existence of 1-Hamiltonian surfaces is now decided for all regular polytopes. Furthermore, 2-Hamiltonian 4-manifolds in the d-dimensional cross polytope are investigated. These are the "regular cases" satisfying equality in Sparla's inequality. In particular, a new example with 16 vertices which is highly symmetric with an automorphism group of order 128 is presented. Topologically, it is homeomorphic to a connected sum of 7 copies of $S^2 \times S^2$. By this example all regular cases of n vertices with $n < 20$ or, equivalently, all cases of regular d-polytopes with $d \leq 9$ are now decided.

The notion of *tightness* of a PL-embedding of a triangulated manifold is closely related to its property of being a Hamiltonian subcomplex of some convex polytope. Tightness of a triangulated manifold is a topological condition, roughly meaning that any simplex-wise linear embedding of the triangulation into Euclidean space is "as convex as possible". It can thus be understood as a generalization of the concept of convexity. In even dimensions, super-neighborliness is known to be a purely combinatorial condition which implies the tightness of a triangulation. Here, we present other sufficient and purely combinatorial conditions which can be applied to the odd-dimensional case as well. One of the conditions is that all vertex links are stacked spheres, which implies that the triangulation is in Walkup's class $\mathcal{K}(d)$. We show that in any dimension $d \geq 4$ *tight-neighborly* triangulations as defined by Lutz, Sulanke and Swartz are tight. Also, triangulations with k-stacked vertex links and the centrally symmetric case are discussed.

Furthermore, a construction of an infinite series of simplicial complexes M^{2k} in rising, even dimensions is presented. It is conjectured that for each k, (i) M^{2k} is a centrally symmetric triangulation of a sphere product $S^k \times S^k$, which (ii) is a k-Hamiltonian subcomplex of the $(2k+2)$-cross polytope, that (iii) satisfies equality in Sparla's inequality. Using the software package `simpcomp`, it is shown that the conjecture holds for $k \leq 12$.

2000 MSC classification: 52B05, 52B70, 53C42, 52B70, 57Q35.

Key words: Hamiltonian subcomplex, triangulated manifold, sphere products, pinched surface, centrally-symmetric, tight, perfect Morse function, stacked polytope.

Zusammenfassung

Diese Dissertationsschrift ist im Bereich der kombinatorischen Topologie angesiedelt, einem Schnittbereich der Topologie, der Polytoptheorie und der Kombinatorik. Während man in den Anfangstagen der Topologie Mannigfaltigkeiten und deren topologische Invarianten meist anhand der kombinatorischen Struktur ihrer Triangulierungen untersuchte (siehe z.b. die Lehrbücher [119] oder [115]), sich die Berechnung der Invarianten über diesen Weg aber als sehr mühsam herausstellte, strebte man danach, die Invarianten einer Mannigfaltigkeit ohne den Umweg über eine Triangulierung zu berechnen. So wurden im Laufe der 1930er und 1940er Jahre die kombinatorischen Methoden Stück für Stück durch solche algebraischer Natur abgelöst.

Seit dem Aufkommen von Computern wurde aber den kombinatorischen Aspekten von Mannigfaltigkeiten und ihren Triangulierungen erneut großes Interesse zuteil, da nun die von Hand mühselige Arbeit des Umgangs mit großen Triangulierungen nichttrivialer Objekte und der Berechnung von deren Invarianten durch Verwendung geeigneter Computerprogramme (siehe [51, 44, 45, 52]) dem Rechner überlassen werden kann. Diese Entwicklung zeigt auch die große Zahl an Publikationen in diesem Bereich während der letzten Jahre, siehe [92] für eine Übersicht.

In dieser Arbeit werden Hamiltonsche Untermannigfaltigkeiten konvexer Polytope und ihre Eigenschaften untersucht. Eine Untermannigfaltigkeit M eines polyedrischen Komplexes K heißt k-Hamiltonsch, wenn M das ganze k-Skelett von K enthält. Diese Verallgemeinerung der Idee eines Hamiltonschen Zyklus in einem Graphen geht zurück auf die Arbeiten [48] und [117]. Ist hier K der Randkomplex eines konvexen Polytops, so wird die Betrachtung auch geometrisch besonders interessant (siehe [78, Kap. 3]). Beispielsweise wurden in [3] geschlossene 1-Hamiltonsche

Untermannigfaltigkeiten in bestimmten Polytopen untersucht und es gibt berühmte auf Harold Coxeter zurückgehende Beispiele von quadrangulierten Flächen, welche als 1-Hamiltonsche Untermannigfaltigkeiten höherdimensionaler Würfel angesehen werden können, siehe [85]. Eine Übersicht über das Thema von Hamiltonschen Untermannigfaltigkeiten konvexer Polytope findet sich in [77]. Während Existenz und Klassifikation von Hamiltonzykeln in den Skeletten der regulären konvexen 3-Polytope seit langem mathematische Folklore sind (der Fall des Ikosaeders als einzig nicht trivialer findet sich in [60]), war dieses Problem für die natürliche Verallgemeinerung in Form von 1-Hamiltonschen Flächen in den Skeletten der regulären konvexen 4-Polytope außer im Fall des Simplex und des Würfels bisher offen. In Kapitel 2 on page 39 dieser Arbeit wird die Nichtexistenz von 1-Hamiltonschen Flächen im 24-Zell, 120-Zell und im 600-Zell gezeigt. Jedoch lässt das 24-Zell sechs Isomorphietypen von singulären Flächen mit 4, 6, 8 bzw. 10 pinch points zu, welche klassifiziert werden (ein *pinch point* ist eine singuläre Ecke v der Fläche, für welche $\mathrm{lk}(v) \cong S^1 \sqcup S^1$ gilt). Im Fall des 120-Zells konnte auch die Existenz solcher kombinatorischer Pseudomannigfaltigkeiten widerlegt werden, während diese Frage für den 600-Zell wegen ihrer in diesem Fall hohen Komplexität noch nicht entschieden werden konnte. Die Kapitel 4 on page 75 und 5 on page 97 beschäftigen sich mit dem zentralsymmetrischen Fall von Hamiltonschen Untermannigfaltigkeiten im d-dimensionalen Kreuzpolytop (der Verallgemeinerung des Oktaeders). Obwohl es einige theoretische Resultate in dem Gebiet gibt (unter anderem durch Arbeiten von Eric Sparla [124] und Frank Lutz [90]), mangelt es doch an nicht-trivialen Beispielen, die beispielsweise die Schärfe von gewissen Ungleichungen zeigen können. Obwohl der Beweis der Existenz einer vermuteten Serie von triangulierten Sphärenprodukten $S^{k-1} \times S^{k-1}$ im Kreuzpolytop in seiner vollen Allgemeinheit hier weiter schuldig geblieben werden muss, sind in Kapitel 5 zumindest Teilergebnisse und ein Beweis der Existenz der Triangulierungen für $k \leq 12$ aufgeführt. Diese Serie würde eine in [90, Kap. 4.2, S. 85] aufgestellte Vermutung beweisen und die Schärfe einer in [124, Kap. 3] aufgestellten Ungleichung in beliebiger Dimension zeigen.

Die Eigenschaft eines Simplizialkomplexes, eine Hamiltonsche Untermannigfaltigkeit in einem Polytop zu sein, ist eng mit der Eigenschaft seiner "Straffheit" verbunden, d.h. der Eigenschaft, dass alle PL-Einbettungen des Komplexes in einen euklidischen Raum "so konvex wie möglich" sind. Im Fall von Untermannigfaltigkeiten des Simplex spricht man auch von *straffen Triangulierungen*. Straffheit ist in diesem Sinne eine Verallgemeinerung des Begriffs der Konvexität, der nicht nur durch topologische Bälle und deren Randmannigfaltigkeiten erfüllt werden kann. Zum Begriff der Straffheit siehe [78] und zu einer Übersicht bekannter straffer Triangulierungen [84]. Kapitel 3 dieser Arbeit befasst sich mit der Untersuchung einer speziellen Klasse von triangulierten Mannigfaltigkeiten (nämlich solcher, die in Walkups Klasse $\mathcal{K}(d)$ liegen, d.h. solcher, deren Eckenfiguren samt und sonders gestapelte Sphären sind) und leitet für diese Mannigfaltigkeiten kombinatorische Bedingungen her, welche die Straffheit ihrer PL-Einbettungen implizieren. Dies ist außerdem die erste bekannte rein kombinatorische Bedingung, welche die Straffheit einer Triangulierung einer ganzen Klasse von Mannigfaltigkeiten auch in ungeraden Dimensionen $d \geq 5$ impliziert.

Wie bereits oben beschrieben, spielte der Computer bei der Untersuchung und der Erzeugung von den in dieser Arbeit untersuchten Objekten eine entscheidende Rolle. In Kooperation mit meinem Kollegen Dipl.-Math. Jonathan Spreer entwickelte ich deshalb ein Erweiterungspaket zum Softwaresystem GAP, welches die Konstruktion und Untersuchung von simplizialen Komplexen im GAP-System ermöglicht und welches wir auf den Namen simpcomp [44] tauften. simpcomp ist inzwischen schon recht umfangreich und erfreut sich in der GAP-Gemeinschaft anscheinend einer gewissen Beliebtheit. Das Programmpaket ist in Anhang C in seinen Grundzügen beschrieben.

Introduction

In fall 2007 I became a PhD student of Wolfgang Kühnel at the University of Stuttgart. Back then I had already worked some years under his supervision as a research assistant for the DFG project Ku 1203/5. Working for the project titled "Automorphism groups in combinatorial topology" aroused my interest in the field. Soon after I was employed by Wolfgang Kühnel and Wolfgang Kimmerle at the successor DFG-granted project Ku 1203/5-2 here in Stuttgart. Most of the results presented in the work at hand were achieved during my employments for the two projects.

This work is set in the field of *combinatorial topology* (sometimes also referred to as *discrete geometric topology*), a field of research in the intersection of topology, geometry, polytope theory and combinatorics. The main objects of interest in the field are simplicial complexes that carry some additional structure, forming *combinatorial triangulations* of the underlying PL manifolds.

From the first days, combinatorial methods were used in (algebraic) topology. Although the interest in combinatorial decompositions of manifolds in form of simplicial complexes declined for some time when the (with regard to cell numbers) more efficient cell decompositions of manifolds were discovered, they again gained popularity in the topological community with the beginning of the digital age. Now the tedious task of working with large triangulations of nontrivial objects by hand could be delegated to a computer; additionally the structure of simplicial complexes is very well suited for a digital representation and the algorithmic investigation with the help of a computer.

Some typical questions researched in the field include for example: (i) Upper and lower bounds on vertex and face numbers, i.e. the investigation of upper and lower bounds on the number of higher-dimensional faces w.r.t. the number of vertices of a simplicial polytope or more generally a triangulation of a manifold (possibly including additional variables, such as Betti numbers), (ii) Minimal triangulations, i.e. the question of the minimal number of vertices needed for a combinatorial triangulation of a triangulable topological manifold M of a given topological type, (iii) Questions of existence of combinatorial triangulations of some given topological manifold, (iv) Questions relating purely combinatorial properties of the triangulations to geometrical properties of their embeddings.

In this work, the focus is on the last two points given above. Specifically, the question of the existence and the investigation of the properties of so-called *Hamiltonian submanifolds* in certain polytopes will be of interest in the following. Here, a k-Hamiltonian submanifold of a polytope is a submanifold that contains the full k-skeleton of the polytope. Since the case of Hamiltonian submanifolds of the cube is well known and since the case of a simplex was also previously studied a focus is given on the case of the cross polytope and the sporadic regular 4-polytopes: In Chapter 2 the existence of so-called *Hamiltonian surfaces* in the regular convex 4-polytopes is investigated (these are surfaces that contain the full 1-skeleton of the polytope). Surprisingly, it turned out that neither the 24-cell, the 120-cell, nor the 600-cell admit such Hamiltonian surfaces in their boundary complexes. By our results the existence of 1-Hamiltonian surfaces is now decided for all regular polytopes.

The property of a combinatorial submanifold of being a Hamiltonian subcomplex of some higher-dimensional polytope is closely related to a property of PL embeddings of combinatorial manifolds referred to as *tightness*. Roughly speaking, tightness is a generalization of the notion of convexity in the sense that a manifold is "as convex as its topology lets it be" if it is tight, i.e. tightness can be understood as a notion of convexity that also applies to objects other than topological balls and their boundary manifolds. This relation (stemming from the field of differential geometry) was studied extensively among others by Thomas Banchoff, Nicolaas Kuiper and Wolfgang Kühnel. Chapters 3 and 4 investigate properties of triangulations related to their tightness. Chapter 3 contains the discussion of the conditions for the tightness of members of a certain class of triangulated manifolds, namely manifolds in Walkup's class $\mathcal{K}(d)$, i.e. manifolds that have stacked vertex links. For this class a purely combinatorial condition implying tightness of the embedding is given. This condition holds in arbitrary (also odd) dimension $d \geq 4$ and seems to be the first such condition for odd dimensions.

Chapter 4 investigates in greater generality on Hamiltonian submanifolds of cross polytopes, i.e. the centrally symmetric case of Hamiltonian submanifolds and conditions for the tightness of such triangulations.

Chapter 5 contains the construction of a conjectured series of centrally symmetric triangulations of sphere products $S^k \times S^k$ as Hamiltonian subcomplexes of higher-dimensional cross polytopes. Although firmly believed to be true by the author, the statement is unfortunately still a conjecture as of the time being. None the less the findings during the research on the problem and partial results are written down.

Quite some of the problems of this work have been solved with – or at least were investigated upon with – the help of a computer. The programs used are all written in GAP, the well-known system for discrete computational algebra, and can be found in the appendices. During my time as PhD student I worked in cooperation with Jonathan Spreer at the University of Stuttgart and we developed simpcomp, an extension package to the GAP system that provides a wide range of constructions and tools for simplicial complexes; see Appendix C for a short description of the package and its functionality. If you want to get to know the package in more detail, then there is also an extensive manual available.

Stuttgart, November 2010

Chapter 1

Basics

This chapter contains a brief introduction to the fields of polytope theory and the theory of triangulated manifolds. Furthermore, concepts developed in these fields that will be used throughout this work are discussed, as for example simplicial homology and cohomology, bistellar moves on triangulations, the Dehn-Sommerville equations and polyhedral Morse theory. Additionally, a short tear off of the theory of tight triangulations and an overview of upper and lower bounds for triangulated manifolds will be given.

1.1 Polytopes, triangulations and combinatorial manifolds

Polytopes

Polytopes are fundamental geometric objects that have been studied by generations of mathematicians ever since – the foundations of polytope theory were laid out by Euclid in his Elements [47] who was the first to study the regular convex polytopes in dimension three, the so-called *Platonic solids* (see Figure 1.3 on page 4).

The concept of polytopes seems to date back to the Swiss mathematician Ludwig Schläfli, the term *polytope* seems to be coined by Reinhard Hoppe [62]. After being forgotten for quite a while, it was by the works of Branko Grünbaum [56] that the sleeping beauty polytope theory was revived and since then stood in the focus of modern mathematical research. For an introduction to the field, see for example the books [34, 56, 139] or [97].

Definition 1.1 (convex polytope) *The convex hull P of finitely many points in E^d not lying in a common hyperplane is called* convex d-polytope *– P is sometimes also referred to as V-polytope as it is described by its vertex set. Equivalently, a d-polytope P can be described as the bounded intersection of finitely many closed half spaces in E^d such that the intersection set is of dimension d. In this case P is referred to as H-polytope. The two definitions are equivalent, see [139, Lecture 0].*

Since this work focuses on convex polytopes, we will just write polytope from now on, when actually meaning a convex polytope. Each d-polytope P consists of *faces* and its set of k-faces is referred to as the *k-skeleton of P*.

Definition 1.2 (faces, skeleton)

(*i*) *The intersection of a d-polytope P with a supporting hyperplane $h \subset E^d$ of P is called* k-face *of P if* $\dim(h \cap P) = k$. *A 0-face of P is also-called* vertex, *a 1-face is called* edge *and a $(d-1)$-face is called* facet *of P.*

(*ii*) *For a d-polytope P the k-dimensional skeleton (or k-skeleton) denoted by $\mathrm{skel}_k(P)$ is the set of all i-dimensional faces of P, $i \leq k$. The* face-vector *or f-vector of P counts the number of i-faces of P for all $0 \leq i \leq d$,*

$$f(P) := (f_0, \ldots, f_{d-1}, f_d),$$

where f_i equals the number of i-faces of P. Note that $f_d = 1$ always holds here. In some cases it is of use to formally set

$$f(P) := (f_{-1}, f_0, \ldots, f_{d-1}, f_d)$$

with $f_{-1} := 1$, as the empty set has dimension -1 and is contained in all faces of P.

See Figure 1.1 on the facing page for an illustration of the skeletons of the ordinary 3-cube as convex 3-polytope. For polytopes, the notion of a neighborhood of a vertex can be defined as follows.

Figure 1.1: From left to right: The 0-, 1- and 2-skeleton of the 3-cube.

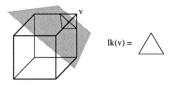

Figure 1.2: A plane separating one vertex v of a 3-cube from all other of its vertices. The intersection of the cube with the plane is the vertex link of v, written $\mathrm{lk}(v)$, in this case a triangle.

Definition 1.3 (vertex figure) *For each vertex v of a convex d-polytope P the vertex figure (or vertex link) of v in P, written as $\mathrm{lk}_P(v)$ or just $\mathrm{lk}(v)$, is defined as the $(d-1)$-polytope which occurs as the intersection of a hyperplane with P separating v from all other vertices of P.*

See Figure 1.2 for an example drawing showing a vertex link in a 3-polytope.

Throughout this work we will most of the time only be interested in the special class of so-called regular polytopes, which can be defined recursively.

Definition 1.4 (regular polytope) *A d-polytope P for which all facets are congruent regular $(d-1)$-polytopes and for which all vertex links are congruent regular $(d-1)$-polytopes is called regular, where the regular 2-polytopes are regular polygons.*

In dimension $d = 2$, every convex regular polygon is a regular polytope and hence there exist infinitely many regular polytopes in this dimension. In dimension $d = 3$, there exist five regular convex polytopes, the so-called Platonic solids, see Figure 1.3 on the following page and Section 2.1 on page 40. In dimension $d = 4$, there exist six regular convex polytopes, the 4-simplex, the 4-cube and its dual the 4-octahedron, the 24-cell and the 120-cell and its dual the 600-cell, see Section 2.2 on page 42. In dimensions $d \geq 5$ the only regular convex polytopes are the d-simplex

Figure 1.3: The five regular convex 3-polytopes, from left to right: the tetrahedron, the cube and its dual the octahedron, the dodecahedron and its dual the icosahedron.

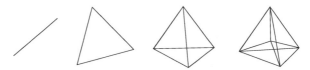

Figure 1.4: From left to right: the (-1)-simplex (the empty set), the 0-simplex (a vertex), the 1-simplex (a line segment), the 2-simplex (a triangle), the 3-simplex (a tetrahedron) and a Schlegel diagram of the 4-simplex, see Section 1.8 on page 36.

(see Definition 1.5), the d-cube and its dual, the d-cross polytope or d-octahedron (see Chapter 4 on page 75).

A special kind of regular polytope – and in fact, the "smallest" regular polytope with respect to the number of vertices in any given dimension – is the *simplex*. See Figure 1.4 for a visualization of some simplices of small dimensions.

Definition 1.5 (simplex, simplicial and simple polytopes) *The d-simplex Δ^d is the convex hull of $d+1$ points in general position in E^d. Δ^d has $\binom{d+1}{i+1}$ i-faces. A d-polytope P is called* simplicial *if for any $i < d$, each of its i-faces is an i-simplex. Here it suffices to ask that all facets of P are simplices. P is called* simple *if its dual is simplicial.*

The d-simplex Δ^d has another specialty: it is the only d-polytope for which any tuple of vertices is the vertex set of a face of Δ^d. Thus, it is said to be $(d+1)$-neighborly.

Definition 1.6 (neighborliness, neighborly polytope) *A d-polytope P is called k-neighborly if any tuple of k or less vertices is the vertex set of a face of P. A $\lfloor\frac{d}{2}\rfloor$-neighborly polytope is called* neighborly polytope *as no d-polytope other than the d-simplex Δ^d can be more than $\lfloor\frac{d}{2}\rfloor$-neighborly.*

Note that any neighborly polytope is necessarily simplicial.

Polytopal and simplicial complexes

Each d-polytope P can be assigned its *polytopal complex* and *boundary complex* as defined below. See Figure 1.5 on the following page for a visualization of the boundary complex of the 3-cube.

Definition 1.7 (polytopal complex, [139])

(i) A polytopal complex C *is a finite collection of convex polytopes in E^d (called facets) that satisfies the following conditions:*

a) *C contains the empty polytope $\varnothing \in C$,*

b) *if $P \in C$, then C also contains all faces of P,*

c) *if $P, Q \in C$, then $P \cap Q$ is either empty or a common face of P and Q.*

The dimension of C is the maximal dimension of a facet of C. If all facets of C have the same dimension, C is called pure. The k-skeleton and the f-vector of a polytopal complex is explained in the same way as for polytopes.

(ii) *The underlying set $|C|$ of C is the union of all polytopes of C as a point set in E^d:*

$$|C| := \bigcup_{P_i \in C} P_i \subset E^d.$$

(iii) *Given a d-polytope P, the set*

$$C(P) := \bigcup_{i=0}^{d} \mathrm{skel}_i(P)$$

of all faces of P is a polytopal complex, the polytopal complex of P. The set of all proper faces of P also forms a polytopal complex,

$$C(\partial P) := \bigcup_{i=0}^{d-1} \mathrm{skel}_i(P).$$

It is called the boundary complex $C(\partial P)$ of P.

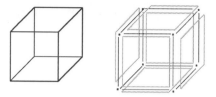

Figure 1.5: The 3-cube (left) and its boundary complex (right), where the faces of the different dimensions are drawn in different shades of gray.

Figure 1.6: The vertex star st(v) (drawn shaded) and the vertex link lk(v) (drawn as thick line) of the vertices v, w in two 2-dimensional complexes.

(iv) Let C be a polytopal complex and $D \subset C$ such that D is a polytopal complex. Then D is called (polytopal) subcomplex *of D.*

In order to establish the notion of a neighborhood in polytopal complexes, one can define the *star* and the *link* of faces in polytopal complexes.

Definition 1.8 (star and link, see [139]) *Let σ be a face of some polytopal complex C. Then the* star *of σ in C is defined as the polytopal complex of facets of C that contain σ as a face, and their faces:*

$$\mathrm{st}_C(\sigma) := \{\tau \in C : \exists_{P \in C} : \sigma \subset P, \tau \text{ face of } P\}.$$

The link *of σ in C then is*

$$\mathrm{lk}_C(\sigma) := \{\tau \in \mathrm{st}_C(\sigma) : \sigma \cap \tau = \varnothing\}.$$

Whenever it is clear what the ambient complex is, we will not write $\mathrm{lk}_C(\sigma)$ *and* $\mathrm{st}_C(\sigma)$*, but just* $\mathrm{lk}(\sigma)$ *and* $\mathrm{st}(\sigma)$*, respectively.*

See Figure 1.6 for an illustration of the star and the link of vertices in two 2-dimensional complexes.

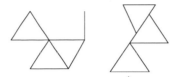

Figure 1.7: Two collections of simplices, a (non-pure) 2-dimensional simplicial complex on the left and a collection of 2-simplices that is not a simplicial complex on the right.

Note that for the boundary complex $C(\partial P)$ of a simplicial polytope P, the two notions of a vertex figure in P (see Definition 1.3 on page 3) and the link of the corresponding vertex in $C(\partial P)$ (see Definition 1.8) coincide, cf. Figure 1.6 (left), whereas in general this does not hold for arbitrary polytopal complexes, cf. Figure 1.6 (right). In the latter case the corresponding vertex figure in the polytope would be a quadrangle and not an 8-gon. If one were to define a notion that generalizes the vertex figure also for non-simplicial polytopes, one would have to define the star and link as in [46]. Keep in mind though, that the definition given in [46] is not compatible to the one used in this work.

In what follows, we will work with a special class of polytopal complexes most of the time.

Definition 1.9 (simplicial complex) *A polytopal complex consisting only of simplices is called* simplicial complex.

See Figure 1.7 for an example of a set of simplices that forms a simplicial complex (left) and for one that does not (right).

As we are in most cases only interested in the topology of a simplicial complex, we will work with a purely combinatorial representation of the complex as defined below.

Definition 1.10 (abstract simplicial complex) *By labeling the vertices of a simplicial complex C with the natural numbers 1 to n one can identify each k-face of C with a set of cardinality $k + 1$. This way a geometrical simplicial complex C can be identified with its so-called* abstract simplicial complex *(or* face lattice*) as a set of finite sets associated with the faces of C. The face lattice carries the structure*

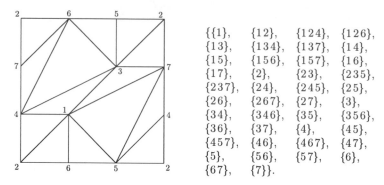

$$\{\{1\}, \quad \{12\}, \quad \{124\}, \quad \{126\},$$
$$\{13\}, \quad \{134\}, \quad \{137\}, \quad \{14\},$$
$$\{15\}, \quad \{156\}, \quad \{157\}, \quad \{16\},$$
$$\{17\}, \quad \{2\}, \quad \{23\}, \quad \{235\},$$
$$\{237\}, \quad \{24\}, \quad \{245\}, \quad \{25\},$$
$$\{26\}, \quad \{267\}, \quad \{27\}, \quad \{3\},$$
$$\{34\}, \quad \{346\}, \quad \{35\}, \quad \{356\},$$
$$\{36\}, \quad \{37\}, \quad \{4\}, \quad \{45\},$$
$$\{457\}, \quad \{46\}, \quad \{467\}, \quad \{47\},$$
$$\{5\}, \quad \{56\}, \quad \{57\}, \quad \{6\},$$
$$\{67\}, \quad \{7\}\}.$$

Figure 1.8: The 7 vertex triangulation of the torus, left: geometrical simplicial complex, right: its face lattice. Vertices with the same labels are identified.

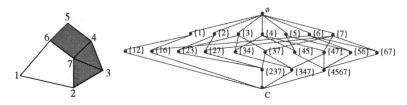

Figure 1.9: A non-pure polytopal complex (left) and its associated Hasse diagram (right).

of a partially ordered set (or poset, for short, see [133]), where the partial order is given by the inclusion.

An abstract simplex on the vertices $v_1, \ldots, v_n$ will be denoted by $\langle v_1 \ldots v_n \rangle$ in the following. See Figure 1.8 for a 7-vertex triangulation T of the torus T^2 as a geometrical simplicial complex (left) and as an abstract simplicial complex in form of its face lattice (right). It is known as *Möbius' Torus* as it was already known to August Möbius [102]. Figure 1.9 shows another example of a polytopal complex and its face lattice visualized via its Hasse diagram.

Two abstract simplicial complexes are considered equal if they are combinatorially isomorphic.

Definition 1.11 (combinatorial equivalence) *Two simplicial complexes C and D are called* combinatorially isomorphic *(or combinatorially equivalent), written*

$C \cong D$, *if their face lattices are isomorphic. This means that there exists a face respecting bijective mapping f of their vertex sets, i.e. a map f that maps faces to faces and if $x, y \in C$, $x \subset y$, then $f(x) \subset f(y)$. A representative of an equivalence class of the equivalence relation $\cong$ represents a* combinatorial type *of simplicial complexes. The set of automorphisms of a simplicial complex C forms a group, the* automorphism group $\mathrm{Aut}(C)$ of C.

There are topological types of complexes for which only one combinatorial type exists for a given fixed number of vertices. These types are called *combinatorially unique*. Möbius' Torus is one example of such a combinatorially unique complex, see [24].

Note that the automorphism group of an abstract simplicial complex on n vertices is always a subgroup of S_n, the symmetric group on n elements. It consists of all face respecting bijective mappings $f : C \to C$.

In the course of this work we will also be interested in a special class of simplicial complexes, so called *centrally symmetric* complexes. These can be characterized on a purely combinatorial level as follows.

Definition 1.12 (abstract central symmetry) *An abstract simplicial complex C is called centrally symmetric if* $\mathrm{Aut}(C)$ *contains an element of order two that acts fixed point free on the face lattice of C.*

The definition generalizes the following geometrical situation: If C can be interpreted as a subcomplex of a convex polytope P such that C contains all vertices of P, then in Definition 1.12 the underlying complex $|C|$ (as subcomplex of P) is centrally symmetric in the usual sense, i.e. there exists a $z \in E^d$ such that

$$x \in |C| \implies 2z - x \in |C|.$$

Triangulated and combinatorial manifolds

In the early days of topology manifolds were triangulated in order to compute topological invariants, see [119].

Definition 1.13 (triangulable and triangulated manifold) *A topological manifold M for which there exists a simplicial complex C such that M is homeomorphic to |C| is called* triangulable manifold. *Any simplicial complex C with |C| ≅ M is referred to as a* triangulation *of M.*

Of course a given triangulable manifold can a priory be triangulated in many different ways. This means that if one wants to compute topological invariants of the underlying manifold using triangulations one has to show that the invariant calculated does not depend on the choice of the triangulation. One such invariant are the homology groups, see Section 1.2 on the next page. Note also that since a simplicial complex may only consist of finitely many simplices, every manifold M that can be triangulated is necessarily compact. Indeed, the converse is also true in low dimensions.

Theorem 1.14 (Rado 1924 & Moise 1954 [103])
For every compact topological manifold M of dimension $d \leq 3$ there exists a combinatorial triangulation of M.

Whether an analogue statement also holds for higher dimensions $d \geq 4$ is not clear as of today.

Since we will work with PL manifolds[1] (for an introduction to PL topology see the books [115] and [63], for more recent developments in the field see [92, 36]) and since the topological and the combinatorial structure need not be compatible in general[2], a slightly stronger notion of a so-called *combinatorial manifold* is introduced here as follows.

Definition 1.15 (combinatorial manifold) *A simplicial complex C that is a triangulation of the topological manifold M is called* combinatorial manifold *of dimension d or* combinatorial triangulation *of M if the link of any i-face of C is a standard PL $(d-i-1)$-sphere. A standard PL $(d-i-1)$-sphere is a simplicial complex which is piecewise linearly homeomorphic to the boundary of the $(d-i)$-simplex $\partial \Delta^{d-i}$.*

---

[1]A PL structure on a manifold is an atlas of charts which are compatible to each other by piecewise linear coordinate transforms.

[2]There exists a triangulation in form of the so-called *Edwards sphere* as a double suspension of a homology 3-sphere which does not carry a PL structure [22].

Figure 1.10: Sketch of a (part of a) surface S transformed into a pinched surface S' – in the polyhedral case S' is obtained from S by subsequent identification of a finite number of vertex pairs of S.

This definition implies that M carries a PL structure. Conversely, every PL manifold admits a triangulation which is a combinatorial manifold in the sense of Definition 1.15 on the facing page. See Figure 1.8 on page 8 for an example of a combinatorial triangulation of the torus.

A *combinatorial d-pseudomanifold* is an abstract, pure simplicial complex M of dimension d such that all vertex links of M are combinatorial $(d-1)$-manifolds in the sense of Definition 1.15. If the vertex link of a vertex v of M is not PL homeomorphic to the $(d-1)$-simplex, that vertex of M is called a *singular vertex* of M.

In the two-dimensional case, a special case of pseudomanifolds are the so called *pinch point* surfaces – here the vertex links are homeomorphic to 1-spheres or disjoint unions of 1-spheres, see Figure 1.10.

1.2 Simplicial homology and cohomology

Why bother triangulating manifolds at all? One of the reasons is to be able to efficiently compute topological invariants of the manifolds via their triangulations. In addition to the powerful, but – apart from the fundamental group π_1 – hard to compute homotopy groups, homology and cohomology groups have proven to be valuable tools for the task of investigating the topological structure of manifolds in terms of algebraic invariants. We will only deal with the simplicial case in the following as we will not need the more general singular theory in this work – but we point out that the constructions are the same in the latter case. For a comprehensive introduction to the subject see the books [105] (the notation of which we will allude to), [123], or [110].

Homology groups

From now on let G be an arbitrary but fixed abelian group, the *group of coefficients*. If not stated otherwise, we will assume that $G = (\mathbb{Z}, +)$. The goal in what follows is to define the "homology groups of a simplicial complex with coefficients in G".

Definition 1.16 (oriented simplex) *Let K be a simplicial complex and let $\varnothing \neq \sigma \in K$ be a face of K. Then an* orientation *of σ is an equivalence class of the arrangements of its vertex set $V(\sigma)$, where two arrangements are considered equivalent if they differ by an even permutation. In the following we will write $[v_1, \ldots, v_k]$ for the orientation induced by the vertex ordering $v_1 < \cdots < v_k$, and $-[v_1, \ldots, v_k]$ for the opposite orientation. Every linear ordering $v_1 < \cdots < v_n$ of the vertex set $V(K)$ of K thus induces an orientation on every face of K.*

Any simplicial complex can be given an ordering of its vertices such that the following construction is well-defined.

Definition 1.17 (simplicial chains and chain group) *Let K be a simplicial complex and let G be a group of coefficients. A q-chain of K with coefficients in G is a formal linear combination*

$$\sum_{\sigma \in \mathrm{skel}_q(K)} \lambda_\sigma \sigma,$$

with $\lambda_\sigma \in G$ and in which every simplex σ may only appear once.

The set of all q-chains of K with coefficients in G forms an additive group (where the addition is defined coefficient-wise for simplices with the same orientation) and is called the q-th chain group *of K with coefficients in G, written $C_q(K; G)$. $C_q(K; G)$ is a free abelian group of rank $f_q(K)$. For $q < 0$ and $q > \dim(K)$ we set $C_q(K; G) := (\{0\}, +)$ and for ease of notation often just write $C_q(K; G) = 0$ from now on.*

We call the sequence of groups $(C_q(K; G))_{q \in \mathbb{Z}}$ the chain complex *$C_*(K; G)$ induced by the complex K.*

Definition 1.18 (boundary map) *The q-th boundary map ∂_q is the group homomorphism $\partial_q : C_q(K;G) \rightarrow C_{q-1}(K;G)$, given by*

$$\partial_q : [v_1, \ldots, v_{q+1}] \mapsto \sum_{i=1}^{q+1} (-1)^{i-1} [v_1, \ldots, \hat{v}_i, \ldots, v_{q+1}].$$

Here the notation $[v_1, \ldots, \hat{v}_i, \ldots, v_{q+1}]$ denotes the $(q-1)$-simplex $[v_1, \ldots, v_{i-1}, v_{i+1}, \ldots, v_{q+1}]$ without the vertex v_i.

Using the boundary map, we can define cycles and boundaries as follows.

Definition 1.19 (cycles and boundaries) *Let K be a simplicial complex and let G be a group of coefficients. The q-th cycle group of K with coefficients in G is the group*

$$Z_q(K;G) := \ker(\partial_q) = \{\sigma \in C_q(K;G) : \partial_q(c) = 0\},$$

the q-th boundary group of K with coefficients in G is the group

$$B_q(K;G) := \mathrm{im}(\partial_{q+1}) = \{\partial_{q+1}(\sigma) : \sigma \in C_{q+1}(K;G)\}.$$

Note that the groups $Z_q(K;G)$ and $B_q(K;G)$ are subgroups of the free group $C_q(K;G)$ and thus again free.

As the boundary map suffices the identity $\partial_q \circ \partial_{q+1} = 0$ (this follows by explicit calculation), one has $B_q(K;G) \subseteq Z_q(K;G)$ and the following construction is well defined.

Definition 1.20 (homology groups) *Let K be a simplicial complex and let G be a group of coefficients. The q-th homology group of K with coefficients in G is the group*

$$H_q(K;G) := Z_q(K;G)/B_q(K;G).$$

The integer $\beta_q = \mathrm{rank}_G H_q(K;G)$ is called the q-th Betti number of K with respect to the group coefficients G.

For $G = \mathbb{Z}$, the homology groups can be written in the form

$$H_q(K;\mathbb{Z}) = \mathbb{Z}^{\beta_q} \oplus \mathbb{Z}_{t_1} \oplus \cdots \oplus \mathbb{Z}_{t_k}$$

with $t_i | t_{i+1}$ by virtue of the fundamental theorem of finitely generated abelian groups. From now on we will fix $G = \mathbb{Z}$ as our group of coefficients unless stated otherwise.

So far so good. But it remains to show that the homology groups are indeed a topological invariant.

Theorem 1.21

Let K, L be simplicial complexes that are homotopy equivalent and let G be a group of coefficients. Then $H_*(K;G) \cong H_*(L;G)$.

This is shown in a few steps which are sketched in the following. First it is shown that simplicial maps $f : K \rightarrow L$ (i.e. maps f that map simplices of K to simplices of L) induce homomorphisms $f_{\#} : C_*(K;G) \rightarrow C_*(L;G)$ on the chain complexes induced by K and L. These in term induce maps $f_* : H_*(K;G) \rightarrow H_*(L;G)$ on the homology, where chain homotopic maps $f_{\#}$ and $g_{\#}$ induce identical maps f_* and g_* on the homology groups. Now since (i) any continuous map $c : |K| \rightarrow |L|$ can be approximated by a simplicial map on a sufficiently subdivided triangulation, (ii) homotopic maps are chain homotopic after a suitable subdivision and (iii) the induced mappings on the homology are invariant under the process of subdivision, it follows altogether that the simplicial homology groups are homotopy invariants and therefore of course also topologically invariant.

Using homology groups we can define a special class of combinatorial pseudomanifolds, the so called *homology d-manifolds*, as simplicial complexes for which each vertex link has the same homology as the $(d-1)$-simplex, but not necessarily is PL homeomorphic to the $(d-1)$-simplex. *Eulerian d-manifolds* are defined analogously, but here the condition on the vertex links is even weaker, namely that they all have the same Euler characteristic as the $(d-1)$-sphere, as defined below.

Knowing that the simplicial homology groups are homotopy invariants we can define another very important topological invariant of a triangulated manifold, it's *Euler characteristic*.

Definition 1.22 (Euler characteristic) *Let K be a simplicial complex of dimension d and let G be an abelian group of coefficients. The Euler characteristic of K is a topological invariant given by the alternating sum*

$$\chi(K) := \beta_0(K;G) - \beta_1(K;G) + \cdots + (-1)^d \beta_d(K;G) = \sum_{i=0}^{d}(-1)^i \operatorname{rank} H_i(K;G). \quad (1.1)$$

Interestingly, the Euler characteristic of a simplicial complex can be computed without knowing anything about its homology groups.

Theorem 1.23 (Euler-Poincaré formula)

Let K be a simplicial complex of dimension d and G an abelian group of coefficients. Then the following holds:

$$\chi(K) = \sum_{i=0}^{d}(-1)^i f_i(K) = \sum_{i=0}^{d}(-1)^i \beta_i(K;G).$$

Note that on the one hand the f-vector is not a topological invariant (but invariant under the choice of G), whereas on the other hand the Betti numbers are topologically invariant but not invariant under the choice of G.

1.24 Remark *The definition of the Euler characteristic χ can be naturally extended to any topological space for which equation (1.1) remains meaningful, e.g. arbitrary polytopal complexes or more generally topological manifolds that can be triangulated or decomposed into cell complexes.*

Apart from the ordinary homology groups of a complex one is often interested in the relative homology of a pair of complexes $K' \subset K$, i.e. the homology groups of K modulo K'.

Definition 1.25 (relative homology) *Let G be an abelian group, K a simplicial complex and $A \subset K$ a subcomplex. Then the set*

$$C_p(K, A; G) := C_p(K;G)/C_p(A;G)$$

of relative chains of K modulo A with coefficients in G *carries the structure of a free abelian group. As the restriction of the boundary operator $\partial|_A$ is well defined on $C_*(A; G)$, one can define the* relative cycle groups $Z_p(K, A; G)$, *the* relative boundary groups $B_p(K, A; G)$ *and the* relative homology groups $H_p(K, A; G)$ *with coefficients in G analogously as for the ordinary homology groups.*

The natural inclusion map ι and the natural projection map π yield a short exact sequence of chain complexes.

$$0 \; \to \; C_*(A; G) \; \overset{\iota}{\to} \; C_*(K; G) \; \overset{\pi}{\to} \; C_*(K, A; G) \; \to \; 0.$$

The relative homology groups $H_*(K, A; G)$ cannot "see what happens in A", cf. [105, Theorem 27.2].

Theorem 1.26 (Excision theorem)
Let A and U be subspaces of the topological space X such that $U \subset \mathrm{int}(A)$ and both pairs (X, A), $(X \backslash U, A \backslash U)$ are triangulable. Then the natural inclusion map induces an isomorphism

$$H_k(X \backslash U, A \backslash U; G) \cong H_k(X, A; G),$$

where G is an abelian group of coefficients.

Since every short exact sequence of chain complexes induces a long exact sequence of homology groups, we have the following

Theorem 1.27 (long exact sequence for the relative homology)
Let K be a simplicial complex and $A \subseteq K$ be a subcomplex. Then there is a long exact sequence

$$\ldots \; \to \; H_{k+1}(K, A; G) \; \overset{\partial_*}{\to} \; H_k(A; G) \; \overset{\iota_*}{\to} \; H_k(K; G) \; \overset{\pi_*}{\to} \; H_k(K, A; G) \; \overset{\partial_*}{\to} \; H_{k-1}(A; G) \; \overset{\iota_*}{\to} \; \ldots,$$

where G is an abelian group of coefficients.

Using homology groups we can define the notion of an orientation of a triangulated manifold as follows.

Definition 1.28 (orientation) *Let M be a connected d-manifold and let G be an abelian group of coefficients. Then for each point $x \in M$ the relative homology*

$$H_d(M, M\backslash\{x\}; G)$$

is free of rank 1 by excision. A G-orientation of M is the map

$$M \to H_d(M, M\backslash\{x\}; G)$$

that maps each point $x \in M$ to a generator ω_x of $H_d(M, M\backslash\{x\}; G)$, subject to the following condition: for all $x \in M$ there exists a neighborhood U of x and an element $\omega_U \in H_d(M, M\backslash U; G)$ such that for all $y \in U$ the natural map $H_d(M, M\backslash U; G) \to H_d(M, M\backslash\{y\}; G)$ satisfies $\omega_U \mapsto \omega_y$. Such a manifold is also called G-orientable, where the usual choice is $G = \mathbb{Z}$. If M cannot be given an $\mathbb{Z}$-orientation, M is called non-orientable, *otherwise it is said to be* orientable.

Famous examples of non-orientable manifolds are the Möbius strip and the Klein bottle. Note that for $G = \mathbb{Z}_2$ the choice of generator above is unique so that all manifolds are $\mathbb{Z}_2$-orientable.

Orientability also carries over to triangulations of manifolds. In terms of a triangulated manifold M the question of orientability simplifies to the question of existence of a linear ordering of the vertices $v_1, \ldots, v_n$ of M (and its induced orientation $[v_1, \ldots, v_n]$) subject to the condition that each $(d-1)$-face of M must occur in two opposite orientations in the two facets it is contained in.

Cohomology groups

By dualizing the notions developed for homology groups, we obtain the so-called cohomology groups. On the one hand this natural construction has the advantage of carrying more structure than the homology groups (more on that later) while being less geometrically intuitive on the other hand.

Definition 1.29 (cochain complex) *A series C^k of abelian groups together with a series of homomorphisms $d^k : C^k \to C^{k+1}$ with $d^{k+1} \circ d^k = 0$ is called a* cochain complex $C^* = (C^k, d^k)_{k \in \mathbb{Z}}$.

Definition 1.30 (cohomology groups) *Let K be a simplicial complex, G an abelian group and let $C_*(K;\mathbb{Z})$ be the chain complex of the simplicial homology groups of K with integer coefficients. Then $C^* = (C^k, d^k)$ defined by*

$$C^k := \operatorname{Hom}(C_k, G)$$

and $d^k(f) := f \circ \partial_{k+1}$ is a cochain complex and the cohomology groups

$$H^k(K;G) := \ker(d^k)/\operatorname{im}(d^{k-1})$$

of C^* are called simplicial cohomology groups of K with coefficients in G.

Akin to the homology groups, it can be shown that the cohomology groups are topological invariants in the sense that two homotopy equivalent spaces have isomorphic cohomology groups. Generally speaking, homology and cohomology share many properties, as the dual construction already suggests. For certain complexes this becomes particularly apparent.

Theorem 1.31 (Poincaré duality)
Let M be a connected closed triangulable n-(homology-)manifold. Then

$$H_k(M;G) \cong H^{n-k}(M;G)$$

holds for all k and for arbitrary coefficient groups G, if M is orientable and for any M, if $G = \mathbb{Z}_2$.

Here a *homology d-manifold* is a pure simplicial complex of dimension d such that all vertex links have the same homology as the $(d-1)$-sphere.

So why bother studying cohomology groups at all? As already mentioned earlier, the cohomology groups carry more structure than the homology groups. The cohomology groups form a ring, the so-called *cohomology ring*, by virtue of the following product map endowing it with a graded structure. We will use elements of a commutative ring R with unity as our coefficients in the following.

Definition 1.32 (cup product) *Let K be a simplicial complex with a linear ordering of its vertices $v_0 < \cdots < v_n$ and let R be a ring of coefficients which is commutative and has a unity element. The simplicial cup product with coefficients in R is given by the homomorphism*

$$\smile: C^p(K;R) \otimes C^q(K;R) \;\to\; C^{p+q}(K;R),$$

defined by

$$\langle c^p \smile c^q, [v_0, \ldots, v_{p+q}] \rangle = \langle c^p, [v_0, \ldots, v_p] \rangle \cdot \langle c^q, [v_p, \ldots, v_{p+q}] \rangle,$$

if $v_0 < \cdots < v_{p+q}$ in the given ordering and where the operation $\cdot$ on the right hand side denotes the multiplication in R. The cochain $c^p \smile c^q$ is referred to as the cup product *of the cochains c^p and c^q. The map $\smile$ is bilinear and associative and induces a bilinear and associative map*

$$\smile: H^p(K;R) \otimes H^q(K;R) \;\to\; H^{p+q}(K;R),$$

which is independent of the ordering of the vertices of K. The cup product is anti-commutative in the following sense:

$$\alpha^p \smile \beta^q = (-1)^{pq} \beta^p \smile \alpha^q,$$

where $\alpha^p \in H^p(K;R)$ and $\beta^q \in H^q(K;R)$.

Together with the cup product, the external direct sum of all cohomology groups $\bigoplus_{p\geq 0} H^p(K;R)$ is endowed with the structure of a non-commutative but associative ring with unity, the *cohomology ring of K with coefficients in R*.

The Poincaré duality 1.31 on the facing page manifests itself in the cohomology ring, as can be seen in the following result.

Theorem 1.33 (dual pairing)
Let $\mathbb{F}$ be a field and let M be a triangulated, closed, $\mathbb{F}$-orientable d-manifold. Then for each $0 \leq k \leq d$, the cup product induces the following a non-degenerate bilinear

map,

$$\cdot : H^k(M;\mathbb{F}) \otimes H^{d-k}(M;\mathbb{F}) \;\to\; H^d(M;\mathbb{F}) \cong \mathbb{F},$$

For two element $\alpha \in H^k(M;\mathbb{F})$, $\beta \in H^{d-k}(M;\mathbb{F})$, *the product* $\alpha \cdot \beta$ *is referred to as* intersection product *of* α *and* β.

In particular, for an orientable, closed, triangulated manifold M of dimension $4n$, one can define its *intersection form* as follows:

$$
\begin{array}{ccccc}
q_M: & H^{2n}(M;\mathbb{Z}) & \times & H^{2n}(M;\mathbb{Z}) & \to & \mathbb{Z} \\
& (a & , & b) & \mapsto & \langle a \smile b, [M] \rangle
\end{array},
$$

i.e. as the bilinear map that evaluates the cup product of two $2n$-cocycles α and β on the fundamental cycle $[M]$ of the manifold M given by its orientation. Note that the same construction generalizes to the non-orientable case, using $\mathbb{F}_2$-coefficients for the homology groups.

The intersection form q_M carries vital topological information of the manifold M. In particular, Michael Freedman used the intersection form to classify simply-connected topological 4-manifolds, see [49].

1.3   The Dehn-Sommerville equations

The Dehn-Sommerville equations establish relations between the numbers of faces of simplicial polytopes and triangulated manifolds. For simplicial polytopes, they were proved in dimension $d \leq 5$ and conjectured for higher dimensions by Max Dehn [37] in 1905 and finally proved by Duncan Sommerville [122] in 1927. There also exists a version of the Dehn-Sommerville equations for triangulated manifolds given below.

Theorem 1.34 (Dehn-Sommerville equations for manifolds, [78])
Let $f = (f_{-1}, f_0, \ldots, f_{d-1})$ *denote the f-vector of a $(d-1)$-dimensional combinatorial manifold M. Then the following* Dehn-Sommerville equations *hold:*

$$\sum_{i=0}^{d-1}(-1)^i f_i = \chi(M),$$

$$\sum_{i=2j-1}^{d-1}\binom{i+1}{2j-1}f_i = 0 \qquad \text{for } 1 \le j \le \frac{d-1}{2}, \text{ if } d \text{ is odd,} \qquad (1.2)$$

$$\sum_{i=2j}^{d-1}\binom{i+1}{2j}f_i = 0 \qquad \text{for } 1 \le j \le \frac{d-2}{2}, \text{ if } d \text{ is even.}$$

Note that the first equation of the Dehn-Sommerville equations is just the Euler-Poincaré formula of Theorem 1.23 on page 15.

In terms of the h-vector $h = (h_0, \ldots, h_d)$ defined by

$$h_j := \sum_{i=-1}^{j-1}(-1)^{j-i-1}\binom{d-i-1}{j-i-1}f_i$$

the Dehn-Sommerville equations more simply read as

$$h_j - h_{d-j} = \begin{cases} (-1)^{d-j}\binom{d}{j}(\chi(M)-2) & \text{for } d = 2k+1 \text{ and } 0 \le j \le k \\ 0 & \text{for } d = 2k \text{ and } 0 \le j \le k-1 \end{cases}.$$

The Dehn-Sommerville equations can be proved in different ways – the probably most elegant one is due to Peter McMullen [95] using shelling arguments, while there exists also a more direct proof due to Branko Grünbaum [56, Sect. 9.2] by double-counting incidences. The latter proof uses the relation

$$2f_{d-1}(M) = (d+1)f_d(M)$$

that can be obtained readily for any combinatorial d-manifold M as it fulfills the *weak pseudomanifold property*, i.e. that any $(d-1)$-face of M is contained in exactly two facets of M.

1.4 Upper and lower bounds

Some of the most fundamental and, as it turned out, hard questions in polytope theory and the theory of combinatorial manifolds were questions concerning upper and lower bounds on the f-vector of a (simplicial) polytope or a triangulation with respect to the number of vertices.

The following theorem was known as the "Upper bound conjecture" (UBC) for a long time until this "rather frustrating" upper bound problem [56, Sec. 10.1] was solved by McMullen [95] in 1970. It was later extended by Richard Stanley [129] to the more general case of arbitrary simplicial spheres in 1975. In 1998, Isabella Novik [106] showed that the UBC holds for all odd-dimensional simplicial manifolds as well as a few classes of even-dimensional manifolds (namely those with Euler characteristic 2 as well as those with vanishing middle homology). In 2002, Patricia Hersh and Novik [61] furthermore showed that the UBC holds for some classes of odd-dimensional pseudomanifolds with isolated singularities. The classical version reads as follows.

Theorem 1.35 (McMullen's Upper Bound Theorem (UBT))

Let P be a d-polytope with $n = f_0(P)$ vertices. Then for every k it has at most as many k-faces as the corresponding cyclic polytope $C_d(n)$:

$$f_{k-1}(P) \le f_{k-1}(C_d(n)).$$

Here, equality for any k with $\lfloor \frac{d}{2} \rfloor \le k \le d$ implies that P is neighborly and simplicial.

The *cyclic polytope* $C_n(d)$ is the simplicial neighborly d-polytope defined as the convex hull of n subsequent points on the momentum curve $t \mapsto (t, t^d, \ldots, t^d) \subset E^d$. The face structure of $C_d(n)$ is determined by Gale's evenness condition [56, Sect. 4.7] and independent of the choice of points on the momentum curve.

Finding a lower bound of the f-vector of a simplicial d-polytope (and in greater generality, a triangulation of the d-sphere) with respect to the number of vertices turned out to be equally challenging. A lower bound theorem for simplicial polytopes, "one of the more challenging open problems" in polytope theory [56, Sect. 10.2], was conjectured and after some invalid "proofs" (for $d = 4$ among others by M. Brückner, M. Fieldhouse and Grünbaum) finally proved by David Barnette [18, 16], compare [137] for a proof in the cases $d = 4$ and $d = 5$. Barnette [16] also showed that in order to prove the theorem, it suffices to show the inequality for $f_1(P)$. See [23] for a proof based on a shelling argument.

Theorem 1.36 (Barnette's LBT for simplicial polytopes [16])

Let P be a simplicial d-polytope with f-vector $f = f(P)$. Then the following hold:

$$f_j \geq \binom{d}{j} f_0 - \binom{d+1}{j+1} j \quad \text{for all } 1 \leq j \leq d-2 \tag{1.3}$$

$$f_{d-1} \geq (d-1) f_0 - (d+1)(d-2). \tag{1.4}$$

Here equality is attained for any j if and only if P is a stacked polytope.

See Chapter 3 on page 53 for a definition of stacked polytopes. The discussion of the cases of equality was done by Barnette for $j = d - 1$ and by Louis Billera and C.W. Lee [21] for arbitrary values of j. The proof was later extended by Barnette [17] himself, David Walkup [137] ($d = 4$) and by Gil Kalai [68] (all d) to general triangulated $(d-1)$-manifolds, where Kalai's version was the first to include a discussion of the case of equality for all d and j.

In succession to McMullen's "g-conjecture" [96], a combinatorial characterization of all possible f-vectors of simplicial polytopes that subsumes the lower and the upper bound theorem and that was finally proved by Billera and Lee [21] and Stanley [130, 132] in 1979. McMullen and Walkup [98] conjectured a generalized lower bound theorem for simplicial polytopes that was later proved by Stanley [130, 131].

Theorem 1.37 (Generalized LBT, [130, 131])

Let P be a simplicial d-polytope with f-vector $f = f(P)$. Then for $0 \leq j \leq \frac{d-1}{2}$ the following inequality holds:

$$\sum_{i=-1}^{j} (-1)^{j-i} \binom{d-i}{j-i} f_i \geq 0, \tag{1.5}$$

or, equivalently in terms of the h-vector $h = h(P)$ of P:

$$h_{j+1} - h_j \geq 0.$$

Note that for $j = 0$, inequality (1.5) on the previous page is just the trivial inequality

$$f_0 \geq d + 1$$

and for $j = 1$ it is equivalent to Barnette's Lower Bound Theorem, hence the name Generalized Lower Bound Theorem.

The cases of equality of (1.5) were conjectured by McMullen and Walkup [98] to be realized by k-stacked polytopes, see Section 3.4 on page 67. This has been proved in special cases, but the general case is still an open problem as of today.

In the centrally symmetric case, Eric Sparla [125, 124] proved some upper and lower bound theorems, see also Chapter 4 on page 75.

Another interesting type of inequality, the so called Heawood type inequalities, are discussed in Section 1.6 on page 26 as they are closely related to the notion of tightness of a triangulation.

1.5 Bistellar moves

Bistellar moves (or flips) as introduced by Udo Pachner [111] (thus sometimes also referred to as Pachner moves) have proven to be a valuable tool in combinatorial topology.

In order to define bistellar moves we make use of the so called *join* operations for (abstract) simplicial complexes. The join of two simplicial complexes K_1 and K_2, denoted by $K_1 * K_2$ is defined as follows.

$$K_1 * K_2 := \{\sigma_1 \cup \sigma_2 : \sigma_1 \in K_1, \sigma_2 \in K_2\}.$$

For example we obtain the d-simplex Δ^d by forming the join of Δ^{d-1} with a new vertex not contained in Δ^{d-1}. Let us now come to the definition of bistellar moves.

Definition 1.38 (bistellar moves) *Let M be a triangulated d-manifold and let A be a $(d-i)$-face of M, $0 \leq i \leq d$, such that there exists an i-simplex B that is not* *a face of M with* $\mathrm{lk}_M(A) = \partial B$. *Then a* bistellar *i*-move Φ_A *on M is defined by*

$$\Phi_A(M) := (M\backslash(A * \partial B)) \cup (\partial A * B),$$

where $*$ *denotes the join operation for simplicial complexes. Bistellar i-moves with* $i > \lfloor \frac{d}{2} \rfloor$ *are also-called* reverse $(d-i)$-moves.

Note that a bistellar 0-move is nothing else than a stellar subdivision of a facet and that an i-move and the corresponding reverse i-move cancel out each other.

Bistellar moves can be used to define an equivalence relation on the set of all pure simplicial complexes.

Definition 1.39 (bistellar equivalence) *We call two pure simplicial complexes* bistellarly equivalent, *if there exists a finite sequence of bistellar moves transforming one complex into the other.*

As it turns out, this indeed is an equivalence relation, also from the topological point of view.

Theorem 1.40 (Pachner [111])
Two combinatorial manifolds are PL homeomorphic if and only if they are bistellarly equivalent.

Thus, bistellar flips leave the PL homeomorphism type of a given triangulated manifold M invariant. Each flip can be thought of as an edge in the bistellar flip graph, where the vertices of that graph are represented by combinatorial triangulations of $|M|$.

Using this process in a simulated annealing type strategy as was done by Frank H. Lutz and Anders Björner in [22] one can try to obtain small and in some cases even vertex minimal triangulations of some given manifold, compare [91, 90, 92, 93, 89].

For a visualization of bistellar moves in the 3-dimensional case see Figure 1.11 on the next page.

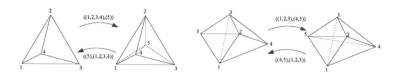

Figure 1.11: Bistellar moves in dimension $d = 3$, left: bistellar 0-move and its inverse 3-move, right: 1-move and its inverse 2-move.

1.6 Tightness, tautness and Heawood inequalities

Tightness is a notion developed in the field of differential geometry as the equality of the (normalized) *total absolute curvature* of a submanifold with the lower bound *sum of the Betti numbers* [87, 14]. It was first studied by Alexandrov [2], Milnor [101], Chern and Lashof [32] and Kuiper [86] and later extended to the polyhedral case by Banchoff [12], Kuiper [87] and Kühnel [78].

From a geometrical point of view, tightness can be understood as a generalization of the concept of convexity that applies to objects other than topological balls and their boundary manifolds since it roughly means that an embedding of a submanifold is "as convex as possible" according to its topology. The usual definition is the following.

Definition 1.41 (tightness [87, 78]) *Let $\mathbb{F}$ be a field. An embedding $M \to \mathbb{E}^k$ of a triangulated compact manifold is called k-tight with respect to $\mathbb{F}$, if for any open or closed half-space $h \subset \mathbb{E}^k$ the induced homomorphism*

$$H_i(M \cap h; \mathbb{F}) \longrightarrow H_i(M; \mathbb{F})$$

is injective for all $i \leq k$. M is called $\mathbb{F}$-tight if it is k-tight with respect to $\mathbb{F}$ for all k. The standard choice for the field of coefficients is $\mathbb{F}_2$ and an $\mathbb{F}_2$-tight embedding is called tight. M is called substantial in E^d if it is not contained in any hyperplane of E^d.

With regard to PL embeddings of PL manifolds, the tightness of a combinatorial manifold can also be defined via a purely combinatorial condition as follows.

Definition 1.42 (tight triangulation) *Let* $\mathbb{F}$ *be a field. A combinatorial manifold* K *on* n *vertices is called* $(k$-$)$*tight w.r.t.* $\mathbb{F}$ *if its canonical embedding* $K \subset \Delta^{n-1} \subset E^{n-1}$ *is* $(k$-$)$*tight w.r.t.* $\mathbb{F}$, *where* Δ^{n-1} *denotes the* $(n-1)$*-dimensional simplex.*

The property of being a tight triangulation is closely related to the so-called Heawood inequality (and its generalizations).

In dimension $d = 2$ the following are equivalent for a triangulated surface S on n vertices: (i) S has a complete edge graph K_n, (ii) S appears as a so-called *regular case* in Heawood's Map Color Theorem 1.43 on the following page (see [59, 114], [78, Chap. 2C]) and (iii) the induced piecewise linear embedding of S into Euclidean $(n-1)$-space has the two-piece property [13], and it is tight [73], [78, Chap. 2D]. Before going to higher dimensions let us discuss the well-understood two-dimensional case a little bit more in detail.

The following inequalities (ii), (iii) and (iv) of Theorem 1.43 are known as *Heawood's inequality* as these were first conjectured by P.J. Heawood [59] in 1890. The problem was solved between 1950 and 1970 by Gerhard Ringel and Ted Youngs [114] for the cases with $g > 0$. For $g = 0$ the still disputed proof of the 4-Color-Problem was accomplished by Appel, Haken and Koch [6, 7] in 1976 with heavily involved, computer-aided proof techniques.

Theorem 1.43 (Map color theorem, G. Ringel, J.W.T. Youngs [114])
Let S *be an abstract surface of genus* g *on* n *vertices which is different from the Klein bottle. The following are equivalent:*

(i) There exists an embedding of the complete graph $K_n \to S$.

(ii) $\chi(S) \leq \dfrac{n(7-n)}{6}$.

(iii) $n \leq \dfrac{1}{2}\left(7 + \sqrt{49 - 24\chi(S)}\right)$.

(iv) $\dbinom{n-3}{2} \leq 3(2 - \chi(s)) = 6g$.

Moreover, equality in the inequalities implies that the embedding of K_n *induces an abstract triangulation of* S *and we will refer to the version*

$$\binom{n-3}{2} \leq 3(2-\chi(s)) = 6g \tag{1.6}$$

as Heawood's inequality *from now on.*

People thus also talk about the uniquely determined *genus of the complete graph* K_n which is (in the orientable regular cases $n \equiv 0, 3, 4, 7$ (12), $n \geq 4$)

$$g = \frac{1}{6}\binom{n-3}{2}.$$

So far so good — but how are the Heawood inequalities related to tight triangulations of surfaces? The following theorem establishes this relation.

Theorem 1.44 (Kühnel, [73])
Let M be an abstract surface and let $n \geq 6$ be a given number. Then the following are equivalent:

(i) There exists a tight and substantial polyhedral embedding $M \rightarrow E^{n-1}$.

(ii) There exists an embedding $K_n \rightarrow M$.

This establishes the correspondence of a tight triangulation of a surface with the case of equality in the Heawood inequality, a so called *regular case* of (1.6), compare [114].

Ringel and Jungerman and Ringel also proved the reversed inequality of (1.6), asking for vertex minimal triangulations of surfaces.

Theorem 1.45 (minimal triangulations of surfaces, [113, 67])
Let M be an abstract surface distinct from the Klein bottle, the orientable surface of genus 2 and from the surface with $\chi = -1$. Then the following are equivalent:

(i) There exists a triangulation of M with n vertices.

(ii) $\binom{n-3}{2} \geq 3(2-\chi(M))$.

Equality in (ii) *above holds if and only if the triangulation is tight.*

They constructed for each case a triangulation of M with the smallest number n of vertices satisfying inequality (ii) on the facing page. Note that the machinery used in the proof is rather involved. In the three exceptional cases the left hand side of the inequality (ii) above has to be replaced by $\binom{n-4}{2}$, compare [64].

After this excursion into the two-dimensional case let us now come to the case of higher dimensions. Here it was Kühnel who investigated the tightness of combinatorial triangulations of manifolds also in higher dimensions and codimensions, see [77], [78, Chap. 4]. It turned out that the tightness of a combinatorial triangulation is closely related to the concept of *Hamiltonicity* of polyhedral complexes (see [76, 78]).

Definition 1.46 (Hamiltonian subcomplex) *A subcomplex A of a polyhedral complex K is called k-Hamiltonian[3] if A contains the full k-dimensional skeleton of K.*

Note that with the simplex as ambient polytope, a k-Hamiltonian subcomplex is a $(k+1)$-neighborly complex, see Definition 1.6 on page 4.

This generalization of the notion of a Hamiltonian circuit in a graph seems to be due to Christoph Schulz [117, 118]. A Hamiltonian circuit then becomes a special case of a 0-Hamiltonian subcomplex of a 1-dimensional graph or of a higher-dimensional complex [48]. See Figure 2.3 on page 46 for the topologically unique Hamiltonian cycles in the tetrahedron and the cube.

If K is the boundary complex of a convex polytope, then this concept becomes particularly interesting and quite geometrical [78, Ch.3]. Amos Altshuler [3] investigated 1-Hamiltonian closed surfaces in special polytopes.

A triangulated $2k$-manifold that is a k-Hamiltonian subcomplex of the boundary complex of some higher dimensional simplex is a tight triangulation as Kühnel [78, Chap. 4] showed.

Theorem 1.47 (Kühnel [78])
Assume that $M \subset P \subset E^d$ is a subcomplex of a convex d-polytope P such that M contains all vertices of P and assume that the underlying set of M is homeomorphic to a $(k-1)$-connected $2k$-manifold. Then the following are equivalent:

---

[3]Not to be confused with the notion of a k-Hamiltonian graph [31].

(i) M is tight in E^d.

(ii) M is k-Hamiltonian in P.

Remember that in the case of the simplex as ambient polytope, k-Hamiltonian subcomplexes are $(k + 1)$-neighborly triangulations. Such $(k + 1)$-neighborly triangulations of $2k$-manifolds are also referred to as *super-neighborly* triangulations – in analogy with neighborly polytopes the boundary complex of a $(2k + 1)$-polytope can be at most k-neighborly unless it is a simplex. Notice here that combinatorial $2k$-manifolds can go beyond k-neighborliness, depending on their topology.

The notion of a *missing face* plays an important role for the tightness of a triangulation. For any tight subcomplex K of the boundary complex of a convex polytope P the following is a direct consequence of Definition 1.41 on page 26, compare [78, 1.4].

Consequence 1.48

A facet of the polytope P is either contained in K or its intersection with K represents a subset of K (often called a topset) which injects into K at the homology level and which is again tightly embedded into the ambient space. In particular, any missing $(k + 1)$-simplex in a k-Hamiltonian subcomplex K of a simplicial polytope represents a non-vanishing element of the k-th homology by the standard triangulation of the k-sphere.

In even dimensions, generalized Heawood inequalities can be obtained between the dimension of the polytope and the Euler characteristic of the manifold as in this case the Euler characteristic $\chi(M) = 2 + (-1)^k \beta_k(M)$ contains essential information about the topology of M. Here β_k denotes the k-th Betti number.

Theorem 1.49 (Kühnel, [77])

Let P be a simplicial d-polytope and let $M \subset P \subset E^d$ be a $(k - 1)$-connected combinatorial $2k$-manifold that is a tight subcomplex of P and that contains all vertices of P. Then the following holds:

$$\binom{d - k - 1}{k + 1} \leq (-1)^k \binom{2k + 1}{k + 2} (\chi(M) - 2) = \binom{2k + 1}{k + 1} \beta_k(M). \qquad (1.7)$$

Moreover, for $d \geq 2k+2$ equality holds if and only if P is a simplex (and, consequently, if M is a tight triangulation).

This can be called "generalized Heawood inequality" as when P is a simplex and $k = 1$, inequality (1.7) on the facing page reads as

$$\binom{d-2}{2} \leq 3(2 - \chi(M)),$$

which is the Heawood inequality with $n = d + 1$, see Theorem 1.43 on page 28.

As in the two-dimensional case, a reverse inequality to (1.7) holds as follows.

Theorem 1.50 (Kühnel, [78])

Assume that M is a $(k-2)$-tight combinatorial triangulation of a $(k-2)$-connected $2k$-manifold with n vertices. Then the following holds:

$$\binom{n-k-2}{k+1} \geq (-1)^k \binom{2k+1}{k+1}(\chi(M) - 2), \tag{1.8}$$

with equality if and only if M is $(k-1)$-connected and the triangulation is tight.

Furthermore, Kühnel [77, 78] conjectured Theorem 1.50 to hold in greater generality for any n-vertex triangulation of a $2k$-manifold, what was later almost proved by Novik [106] and finally proved by Novik and Swartz [108].

Theorem 1.51 (Kühnel's conjecture, [77, 78, 106, 108])

Let M be a combinatorial n vertex triangulation of a $2k$-manifold. Then the following inequality holds:

$$\binom{n-k-2}{k+1} \geq (-1)^k \binom{2k+1}{k+1}(\chi(M) - 2).$$

Equality holds if and only if M is a tight triangulation.

Equality holds precisely in the case of super-neighborly triangulations. These are k-Hamiltonian in the $(n-1)$-dimensional simplex. In the case of 4-manifolds (i.e., $k = 2$) an elementary proof was already contained in [78, 4B].

Except for the trivial case of the boundary of a simplex itself there are only a finite number of known examples of super-neighborly triangulations, reviewed in

[84]. They are necessarily tight (cf. [78, Ch.4]). The most significant ones are the unique 9-vertex triangulation of the complex projective plane [81], [82], a 16-vertex triangulation of a K3 surface [29] and several 15-vertex triangulations of an 8-manifold "like the quaternionic projective plane" [28]. There is also an asymmetric 13-vertex triangulation of $S^3 \times S^3$, see [84], but most of the examples are highly symmetric.

Note that for odd $d = 2k + 1$, inequality (1.7) on page 30 holds trivially, but no conclusion about the case of equality is possible as the boundary of any $(2k + 1)$-polytope is an example. For fixed d, the right hand side of (1.7) gives the minimal "genus" (as the minimal number of copies of $S^k \times S^k$ needed) of a $2k$-manifold admitting an embedding of the complete k-skeleton of the d-simplex. As in the 2-dimensional case, the k-Hamiltonian triangulations of $2k$-manifolds here appear as regular cases of the generalized Heawood inequalities.

With the n-cube as ambient polytope, there are famous examples of quadrangulations of surfaces originally due to Harold Coxeter which can be regarded as 1-Hamiltonian subcomplexes of higher-dimensional cubes [85], [78, 2.12]. Accordingly one talks about the *genus of the d-cube* (or rather its edge graph) which is (in the orientable case)

$$g = 2^{d-3}(d - 4) + 1,$$

see [112], [19]. However, in general the genus of a 1-Hamiltonian surface in a convex d-polytope is not uniquely determined, as pointed out in [117, 118]. This uniqueness seems to hold especially for regular polytopes where the regularity allows a computation of the genus by a simple counting argument.

In the cubical case there are higher-dimensional generalizations by Danzer's construction of a *power complex* 2^K for a given simplicial complex K. In particular there are many examples of k-Hamiltonian $2k$-manifolds as subcomplexes of higher-dimensional cubes, see [85]. For obtaining them one just has to start with a neighborly simplicial $(2k-1)$-sphere K. A large number of the associated complexes 2^K are topologically connected sums of copies of $S^k \times S^k$. This seems to be the standard case.

Centrally-symmetric analogues of tight triangulations of surfaces can be regarded as 1-Hamiltonian subcomplexes of cross polytopes or other centrally symmetric polytopes, see [79]. Similarly, we have the *genus of the d-dimensional cross polytope* [66] which is (in the orientable regular cases $d \equiv 0, 1$ (3), $d \geq 3$)

$$g = \frac{1}{3}(d-1)(d-3).$$

There also exist generalized Heawood inequalities for k-Hamiltonian subcomplexes of cross polytopes that were first conjectured by Sparla [126] and almost completely proved by Novik in [107]. The k-Hamiltonian $2k$-submanifolds appearing as regular cases in these inequalities admit a tight embedding into a higher dimensional cross polytope and are also referred to as *nearly $(k + 1)$-neighborly* as they contain all i-simplices, $i \leq k$, not containing one of the diagonals of the cross polytope (i.e. they are "neighborly except for the diagonals of the cross polytope"), see also Chapter 4 on page 75.

For $d = 2$, a regular case of Heawood's inequality corresponds to a triangulation of an abstract surface (cf. [114]). Ringel [113] and Jungerman and Ringel [67] showed that all of the infinitely many regular cases of Heawood's inequality distinct from the Klein bottle do occur. As any such case yields a tight triangulation (see [73]), there are infinitely many tight triangulations of surfaces.

In contrast, in dimensions $d \geq 3$ there only exist a finite number of known examples of tight triangulations (see [84] for a census), apart from the trivial case of the boundary of a simplex and an infinite series of triangulations of sphere bundles over the circle due to Kühnel [78, 5B], [74].

Apart from the homological definition given in Definitions 1.41 on page 26 and 1.42 on page 27, tightness can also be defined in the language of Morse theory in a natural way: On one hand, the total absolute curvature of a smooth immersion X equals the average number of critical points of any non-degenerate height function on X in a suitable normalization. On the other hand, the Morse inequality shows that the normalized total absolute curvature of a compact smooth manifold M is bounded below by the rank of the total homology $H_*(M)$ with respect to any field of coefficients and tightness is equivalent to the case of equality in this bound, see [84]. This will be investigated upon in the following section.

For the similar notion of *tautness* one has to replace half-spaces by balls (or ball complements) and height functions by distance functions in the definitions of tightness, see [30]. This applies only to smooth embeddings. In the polyhedral case it has to be modified as follows.

Definition 1.52 (tautness, suggested in [15]) *A PL-embedding $M \to \mathbb{E}^N$ of a compact manifold with convex faces is called* PL-taut, *if for any open ball (or ball complement) $B \subset \mathbb{E}^N$ the induced homomorphism*

$$H_*(M \cap \mathrm{span}(B_0)) \longrightarrow H_*(M)$$

is injective where B_0 denotes the set of vertices in $M \cap B$, and $\mathrm{span}(B_0)$ refers to the subcomplex in M spanned by those vertices.

Obviously, any PL-taut embedding is also tight (consider very large balls), and a tight PL-embedding is PL-taut provided that it is PL-spherical in the sense that all vertices are contained in a certain Euclidean sphere. It follows that any tight and PL-spherical embedding is also PL-taut [15].

Corollary 1.53

Any tight subcomplex of a higher-dimensional regular simplex, cube or cross polytope is PL-taut.

In particular this implies that the class of PL-taut submanifolds is much richer than the class of smooth taut submanifolds.

1.7   Polyhedral Morse theory

As an extension to classical Morse theory (see [99] for an introduction to the field), Kühnel [75, 78] developed what one might refer to as a *polyhedral Morse theory*. Note that in this theory many, but not all concepts carry over from the smooth to the polyhedral case, see the survey articles [87] and [14] for a comparison of the two cases.

In the polyhedral case *regular simplex-wise linear functions* on combinatorial manifolds, a discrete analog to the Morse functions in classical Morse theory, are defined as follows.

Definition 1.54 (rsl functions, [75, 78]) *Let M be a combinatorial manifold of dimension d. A function $f : M \rightarrow \mathbb{R}$ is called* regular simplex-wise linear *(or* rsl*), if $f(v) \neq f(v')$ for any two vertices $v \neq v'$ of M and f is linear when restricted to any simplex of M.*

Notice that an rsl function is uniquely determined by its value on the set of vertices and that only vertices can be critical points of f in the sense of Morse theory. With this definition at hand one can define critical points and levelsets of these Morse functions as in the classical Morse theory.

Definition 1.55 (critical vertices, [75, 78]) *Let $\mathbb{F}$ be a field, let M be a combinatorial d-manifold and let $f : M \rightarrow \mathbb{R}$ be an rsl-function on M. A vertex $v \in M$ is called* critical of index k and multiplicity m with respect to f, *if*

$$\dim_{\mathbb{F}} H_k(M_v, M_v \backslash \{v\}; \mathbb{F}) = m > 0,$$

where $M_v := \{x \in M : f(x) \leq f(v)\}$ and H_* denotes an appropriate homology theory with coefficients in $\mathbb{F}$. The number of critical points of f of index i (with multiplicity) are

$$\mu_i(f; \mathbb{F}) := \sum_{v \in V(M)} \dim_{\mathbb{F}} H_i(K_v, K_v \backslash \{v\}; \mathbb{F}).$$

In the following chapters we will be particularly interested in special kinds of Morse functions, so-called *polar* Morse functions. This term was coined by Morse, see [104].

Definition 1.56 (polar Morse function) *Let f be a Morse function that only has one critical point of index 0 and of index d for a given (necessarily connected) d-manifold. Then f is called* polar Morse function.

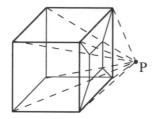

Figure 1.12: Schlegel diagram of the 2-cube (right) obtained by projecting all vertices into one facet of the cube.

Note that for a 2-neighborly combinatorial manifold clearly all rsl functions are polar functions. As in the classical theory, Morse inequalities hold as follows.

Theorem 1.57 (Morse relations, [75, 78])

Let $\mathbb{F}$ be a field, M a combinatorial manifold of dimension d and $f : M \to \mathbb{R}$ an rsl-function on M. Then the following holds, where $\beta_i(M;\mathbb{F}) := \dim_{\mathbb{F}} H_i(M;\mathbb{F})$ denotes the i-th Betti number:

(i) $\mu_i(f;\mathbb{F}) \geq \beta_i(M;\mathbb{F})$ for all i,

(ii) $\sum_{i=0}^{d}(-1)^i \mu_i(f;\mathbb{F}) = \chi(M) = \sum_{i=0}^{d}(-1)^i \beta_i(M;\mathbb{F})$,

(iii) M is (k-)tight with respect to $\mathbb{F}$ if and only if $\mu_i(f;\mathbb{F}) = \beta_i(M;\mathbb{F})$ for every rsl function f and for all $0 \leq i \leq d$ (for all $0 \leq i \leq k$).

Functions satisfying equality in (i) *for all $i \leq k$ are called k-tight functions w.r.t. $\mathbb{F}$. A function f that satisfies equality in* (i) *for all i is usually referred to as $\mathbb{F}$-perfect or $\mathbb{F}$-tight function, cf. [25]. The usual choice of field is $\mathbb{F} = \mathbb{F}_2$.*

Note that a submanifold M of E^d is tight in the sense of Definition 1.41 on page 26 if and only if every Morse function on M is a tight function, see [75, 78].

1.8 Schlegel diagrams

Schlegel diagrams provide a means to visualize a d-polytope in $(d-1)$-dimensional Euclidean space. This tool is especially valuable for the visualization of 4-polytopes, as we will see in Chapter 4 on page 75.

 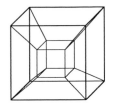

Figure 1.13: Schlegel diagrams of the 4-simplex (left) and the 4-cube (right).

A *Schlegel diagram of a d-polytope P based at the facet F* of P is obtained by a perspective projection of all proper faces of P other than F into F. The projection center x is chosen to lie above the middle of F, i.e. in a plane with ε-distance of and parallel to the supporting hyperplane of P that intersects 'P in F.

This induces a polytopal subdivision of F that can be shown to be combinatorially equivalent to the complex $C(\partial P)\backslash\{F\}$ of all proper faces of P except F.

See Figure 1.12 on the preceding page for a Schlegel diagram of the 3-cube, Figure 1.14 on the following page for examples of Schlegel diagrams of the Platonic solids and Figure 1.13 for examples of Schlegel diagrams of the 4-simplex and the 4-cube.

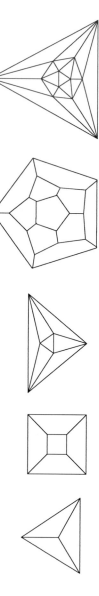

Figure 1.14: Schlegel diagrams of the five Platonic solids as shown in Figure 1.3 on page 4.

Chapter 2

Hamiltonian surfaces in the 24-cell, the 120-cell and the 600-cell

This chapter investigates the question of existence or non-existence of Hamiltonian subcomplexes of certain regular polytopes[1].

It is well-known that there exist Hamiltonian cycles in the 1-skeleton of each of the Platonic solids (see Table 2.1 on page 41 and Figures 2.3 on page 46, 2.7 on page 52). The numbers of distinct Hamiltonian cycles (modulo symmetries of the solid itself) are $1, 1, 2, 1, 17$ for the cases of the tetrahedron, cube, octahedron, dodecahedron, icosahedron, respectively – while the first four cases are easily checked by hand, the more complicated case of the icosahedron was solved by Heinz Heesch in the 1970s, see Figure 2.7 on page 52 and [60, pp. 277 ff.].

Pushing the question one dimension further, a natural question is to ask whether there exist 1-Hamiltonian 2-submanifolds (i.e. Hamiltonian surfaces) in the skeletons of higher dimensional polytopes (cf. [120]). Note here that a 1-Hamiltonian surface in the boundary complex of a Platonic solid must coincide with the boundary itself and is, therefore, not really interesting. Thus, the question becomes interesting only for polytopes of dimension $d \geq 4$.

For $d \geq 5$, the only regular polytopes are the d-simplex which is self-dual, the d-cube and its dual, the d-cross polytope. Since the case of the cube and the simplex were previously studied (see [78, 85, 19]), the focus of attention here will be on the

---

[1]The results of this chapter are in most parts contained in [40], a joint work with Wolfgang Kühnel.

case $d = 4$ (in this chapter) and the case of higher-dimensional cross-polytopes (in Chapter 4 on page 75).

For $d = 4$, Hamiltonian cycles in the regular 4-polytopes are known to exist. However, it seems that so far no decision about the existence or non-existence of 1-Hamiltonian surfaces in the 2-skeleton of any of the three sporadic regular 4-polytopes could be made, compare [120]. This question will be investigated upon in the following.

In this chapter, first the regular convex 3- and 4-polytopes are introduced, followed by an investigation of the question whether there exist 1-Hamiltonian surfaces in the boundary complexes of the four sporadic regular convex 4-polytopes, akin to the equivalent question for 3-polytopes. The answer to this question surprisingly turned out to be negative.

2.1   The five regular and convex 3-polytopes

The five regular convex 3-polytopes as shown in Figure 1.3 on page 4 and Table 2.1 on the facing page – also known as Platonic solids – have been known since antiquity. They were studied extensively by the ancient Greeks, and while some sources credit Pythagoras with their discovery, others account the discovery of the octahedron and icosahedron to Theaetetus, a contemporary of Plato that probably gave the first mathematical proof of their existence along with a proof that there exist no other regular convex 3-polytopes.

Euclid also gave a mathematically complete description of the Platonic solids in his Elements [47]. He used a geometrical proof that there only exist five such polytopes, that is sketched in the following lines:

(i) Each vertex of the polytope is contained in at least three facets and

(ii) at each vertex, the sum of the angles among adjacent facets must be less than 2π.

(iii) Since the geometric situation is the same at each vertex and the minimal vertex number of a facet is 3, each vertex of each facet must contribute an angle less than $\frac{2\pi}{3}$.

Table 2.1: The five Platonic solids and their Schläfli symbols.

name	illustration	Schläfli symbol
tetrahedron		$\{3,3\}$
octahedron		$\{3,4\}$
cube		$\{4,3\}$
dodecahedron		$\{5,3\}$
icosahedron		$\{3,5\}$

(iv) Since regular polygons with six or more sides only admit angles of at least $\frac{2\pi}{3}$ at the vertices, the possible choices for the facets are either triangles, squares or pentagons.

(v) This leaves the following possibilities. For triangular facets: since the angle at each vertex of a regular triangle is $\frac{\pi}{3}$, this leaves the tetrahedron (3 triangles meeting in a vertex), the octahedron (4 triangles meeting in a vertex) and the icosahedron (5 triangles meeting in a vertex) as possibilities. For square facets: since the angle at each vertex is $\frac{\pi}{2}$, this leaves the cube with three squares meeting in a vertex as the only possibility. For pentagonal facets: as the angle at each vertex is $\frac{3\pi}{5}$, again there only exists one solution with three facets meeting at each vertex, the dodecahedron.

Each facet of a Platonic solid is a regular p-gon and the five polytopes can be told apart by p and the number of facets q meeting in a vertex. Consequently, the Platonic solids can be distinguished by their so-called *Schläfli symbol* $\{p,q\}$. See Table 2.1 for a list of the Platonic solids and their Schläfli symbols. The Schläfli symbol reverses its order under dualization: if a regular convex 3-polytope has the Schläfli symbol $\{p,q\}$, then its dual polytope has the Schläfli symbol $\{q,p\}$. This notion can also be generalized to higher dimensions, see Section 2.2 on the following page.

On the other hand a second, combinatorial proof of the fact that there only exist five Platonic solids can be obtained as explained in the following. In fact, the Schläfli symbol $\{p, q\}$ determines all combinatorial information of a regular convex 3-polytope P, as by double counting one gets

$$pf_2(P) = 2f_1(P) = qf_0(P),\tag{2.1}$$

and together with the *Euler relation*

$$f_0(P) - f_1(P) + f_2(P) = 2\tag{2.2}$$

the equations yields

$$f_0(P) = \frac{4p}{4 - (p-2)(q-2)},$$
$$f_1(P) = \frac{2pq}{4 - (p-2)(q-2)},$$
$$f_2(P) = \frac{4q}{4 - (p-2)(q-2)}.$$

Inserting (2.1) into the Euler relation (2.2) yields

$$\frac{2f_1(P)}{q} - f_1(P) + \frac{2f_1(P)}{p} = 2.$$

Consequently,

$$\frac{1}{q} + \frac{1}{p} = \frac{1}{2} + \frac{1}{f_1(P)}, \quad \text{and since } f_1(P) > 0: \quad \frac{1}{q} + \frac{1}{p} > \frac{1}{2}.$$

Since $p \geq 3$ and $q \geq 3$, this only leaves the five possibilities for pairs $\{p, q\}$ listed in Table 2.1 on the previous page.

2.2   The six regular and convex 4-polytopes

As in the three-dimensional case, there exist only a finite number of types —namely six different— of regular convex 4-polytopes, sometimes also referred to as regular

Table 2.2: The six regular convex 4-polytopes.

Name	Schläfli symbol	facets	vert. figures	f-vector
4-simplex	$\{3,3,3\}$	$\{3,3\}$	$\{3,3\}$	$(5,10,10,5)$
4-cube	$\{4,3,3\}$	$\{4,3\}$	$\{3,3\}$	$(16,32,24,8)$
4-cross polytope	$\{3,3,4\}$	$\{3,3\}$	$\{3,4\}$	$(8,24,32,16)$
24-cell	$\{3,4,3\}$	$\{3,4\}$	$\{4,3\}$	$(24,96,96,24)$
120-cell	$\{5,3,3\}$	$\{5,3\}$	$\{3,3\}$	$(600,1200,720,120)$
600-cell	$\{3,3,5\}$	$\{3,3\}$	$\{3,5\}$	$(120,720,1200,600)$

convex *polychora* (Greek choros=room): the 4-simplex, the 4-cube and its dual the 4-octahedron (or 4-cross polytope), the 24-cell (which is self dual) and the 120-cell and its dual, the 600-cell. Table 2.2 lists some properties of these six regular convex 4-polytopes.

Figure 2.1 on the next page shows a visualization of the six polytopes via their Schlegel diagrams (cf. Section 1.8 on page 36), Figure 2.2 on page 45 a second visualization of the polytopes via their 1-skeletons.

2.3 Hamiltonian surfaces in the 24-cell

The boundary complex of the 24-cell $\{3,4,3\}$ consists of 24 vertices, 96 edges, 96 triangles and 24 octahedra. Any 1-Hamiltonian surface (or pinched surface) must have 24 vertices, 96 edges and, consequently, 64 triangles, hence it has Euler characteristic $\chi = -8$.

Every edge in the polytope is in three triangles. Hence we must omit exactly one of them in each case for getting a surface where every edge is in two triangles. Since the vertex figure in the polytope is a cube, each vertex figure in the surface is a Hamiltonian circuit of length 8 in the edge graph of a cube.

It is well known that this circuit is uniquely determined up to symmetries of the cube, see Figure 2.3 on page 46 (left). Starting with one such vertex figure, there are four missing edges in the cube which, therefore, must be in the uniquely determined

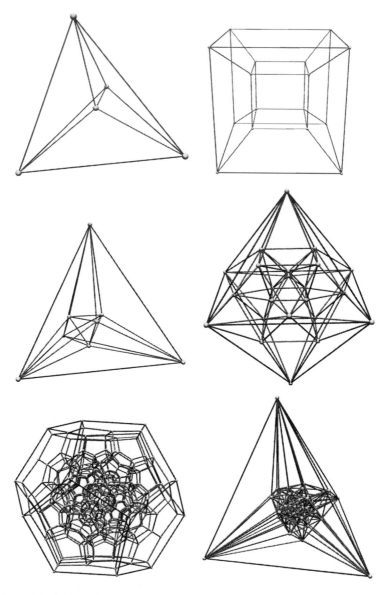

Figure 2.1: Schlegel diagrams of the six regular convex 4-polytopes, from left to right, top to bottom: the 4-simplex, the 4-cube, the 4-octahedron or 4-cross polytope, the 24-cell, the 120-cell and the 600-cell. Visualizations created using the software polymake [52].

Figure 2.2: Visualizations of the six regular convex 4-polytopes that were produced
using the software **jenn** [109]: The 1-skeletons of the polytopes are first
embedded into a 3-sphere and then stereographically projected into
Euclidean 3-space. From left to right, top to bottom: the 4-simplex, the
4-cube, the 4-octahedron or 4-cross polytope, the 24-cell, the 120-cell
and the 600-cell.

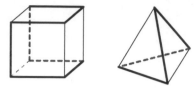

Figure 2.3: A Hamiltonian cycle in the edge graph of the cube (left) and in the edge graph of the tetrahedron (right).

other triangles of the 24-cell. In this way, one can inductively construct an example or, alternatively, verify the non-existence. If singular vertices are allowed, then the only possibility is a link which consists of two circuits of length four each. This leads to the following theorem.

Theorem 2.1

There is no 1-Hamiltonian surface in the 2-skeleton of the 24-cell. However, there are six combinatorial types of strongly connected 1-Hamiltonian pinched surfaces with a number of pinch points ranging between 4 and 10 and with the genus ranging between $g = 3$ and $g = 0$. The case of the highest genus is a surface of genus three with four pinch points. The link of each of the pinch points in any of these types is the union of two circuits of length four.

The six types and their automorphism groups are listed in Tables 2.3 on the next page and 2.4 on the facing page where the labeling of the vertices of the 24-cell coincides with the standard one in the `polymake` system [52]. Visualizations of the six types can be found in [40].

Type 1 is a pinched sphere which is based on a subdivision of the boundary of the rhombidodecahedron, see Figure 2.4 on page 48 (left). Type 4 is just a (4×4)-grid square torus where each square is subdivided by an extra vertex, see Figure 2.4 on page 48 (right). These 16 extra vertices are identified in pairs, leading to the 8 pinch points.

Because -8 equals the Euler characteristic of the original (connected) surface minus the number of pinch points it is clear that we can have at most 10 pinch points unless the surface splits into several components. We present here in more detail Type 6 as a surface of genus three with four pinch points, see Figure 3

Table 2.3: Automorphism groups of the Hamiltonian pinched surfaces in the 24-cell.

type	group	order	generators
1	$C_4 \times C_2$	8	(1 11 12 16 18)(2 17 23 7)(3 13 20 21)(4 22 1 15)(6 19)(8 24 14 10), (1 3)(4 8)(5 10)(9 15)(11 14)(12 13)(16 20)(18 21)(22 24)
2	D_8	8	(1 16)(2 17)(3 22)(5 20)(6 9)(7 23)(8 12)(10 24)(14 18)(15 19), (2 3)(4 6)(5 7)(9 11)(12 14)(13 15)(17 20)(19 21)(22 23)
3	$C_2 \times C_2$	4	(1 24)(2 13)(3 15)(4 17)(5 19)(6 20)(7 21)(9 22)(11 23), (2 5)(3 7)(4 9)(6 11)(8 18)(13 19)(15 21)(17 22)(20 23)
4	$(((C_4 \times C_2) \times C_2) \times C_2) \times C_2$	64	(1 8 10 12)(3 13 5 4)(6 15 19 9)(7 17)(11 20 21 22)(14 24 18 16), (2 3)(4 6)(5 7)(9 11)(12 14)(13 15)(17 20)(19 21)(22 23)
5	S_3	6	(1 3)(4 8)(5 10)(9 15)(11 14)(12 13)(16 20)(18 21)(22 24), (1 22 15)(21 2 13)(3 9 24)(4 17 8)(5 19 10)(6 16 20)(7 18 21)(11 23 14)
6	$C_2 \times D_8$	16	(1 11)(2 23)(3 14)(4 10)(6 19)(7 9)(8 13)(11 18)(14 22)(15 17)(16 21)(20 24), (1 5)(3 12)(4 10)(6 19)(7 9)(8 13)(11 18)(14 22)(15 17)(16 21)(20 24), (1 3)(4 8)(5 10)(9 15)(11 14)(12 13)(16 20)(18 21)(22 24)

Table 2.4: Generating orbits of the 6 types of Hamiltonian pinched surfaces in the 24-cell.

type	# p. pts.	g	orbits
1	10	0	$(123)_4$, $(124)_8$, $(136)_4$, $(149)_8$, $(157)_8$, $(159)_8$, $(1611)_8$, $(1711)_8$, $(2510)_4$, $(468)_4$
2	10	0	$(123)_4$, $(124)_8$, $(149)_4$, $(238)_4$, $(2412)_8$, $(2510)_4$, $(2512)_4$, $(2813)_8$, $(21013)_8$, $(468)_4$, $(81315)_4$, $(101319)_4$
3	8	1	$(123)_4$, $(124)_4$, $(136)_4$, $(149)_2$, $(1611)_2$, $(238)_4$, $(2412)_4$, $(2510)_2$, $(2512)_2$, $(2813)_2$, $(21013)_2$, $(3614)_4$, $(3710)_2$, $(3714)_2$, $(3815)_2$, $(31015)_2$, $(468)_4$, $(4616)_4$, $(4817)_2$, $(4916)_2$, $(41217)_2$, $(6820)_2$, $(61114)_2$, $(61620)_2$
4	8	1	$(123)_{32}$, $(124)_{32}$
5	6	2	$(123)_3$, $(124)_6$, $(136)_3$, $(149)_3$, $(157)_6$, $(159)_3$, $(1611)_6$, $(1711)_6$, $(2412)_3$, $(2510)_3$, $(2512)_3$, $(468)_3$, $(4616)_3$, $(4817)_1$, $(5718)_3$, $(51019)_1$, $(61116)_3$, $(71114)_3$, $(71821)_1$, $(111423)_1$
6	4	3	$(123)_8$, $(124)_8$, $(136)_8$, $(149)_8$, $(157)_{16}$, $(1611)_8$, $(1711)_8$

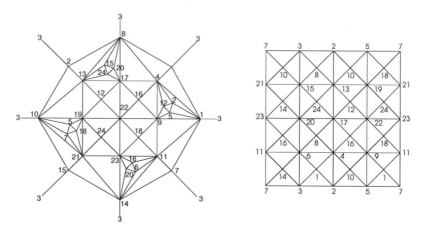

Figure 2.4: Type 1 (left) and Type 4 (right) of Hamiltonian pinched surfaces in the 24-cell.

(produced with `JavaView` [71]). Its combinatorial type is given by the following list of 64 triangles:

$$\begin{array}{lllllll}
(1\,2\,3), & (1\,2\,4), & (1\,3\,6), & (1\,4\,9), & (1\,5\,7), & (1\,5\,9), & (1\,6\,11), & (1\,7\,11), \\
(2\,3\,8), & (2\,4\,8), & (2\,5\,10), & (2\,5\,12), & (2\,10\,13), & (2\,12\,13), & (3\,6\,14), & (3\,7\,10), \\
(3\,7\,14), & (3\,8\,15), & (3\,10\,15), & (4\,6\,8), & (4\,6\,16), & (4\,9\,12), & (4\,12\,17), & (4\,16\,17), \\
(5\,7\,10), & (5\,9\,18), & (5\,12\,19), & (5\,18\,19), & (6\,8\,20), & (6\,11\,14), & (6\,16\,20), & (7\,11\,18), \\
(7\,14\,21), & (7\,18\,21), & (8\,13\,15), & (8\,13\,17), & (8\,17\,20), & (9\,11\,16), & (9\,11\,18), & (9\,12\,22), \\
(9\,16\,22), & (10\,13\,19), & (10\,15\,21), & (10\,19\,21), & (11\,14\,23), & (11\,16\,23), & (12\,13\,17), & (12\,19\,22), \\
(13\,15\,24), & (13\,19\,24), & (14\,15\,20), & (14\,15\,21), & (14\,20\,23), & (15\,20\,24), & (16\,17\,22), & (16\,20\,23), \\
(17\,20\,24), & (17\,22\,24), & (18\,19\,22), & (18\,21\,23), & (18\,22\,23), & (19\,21\,24), & (21\,23\,24), & (22\,23\,24).
\end{array}$$

The pinch points are the vertices 2, 6, 19, 23 with the following links:

$$\begin{array}{rll}
2: & (1\,3\,8\,4) & (5\,10\,13\,12) \\
6: & (1\,3\,14\,11) & (4\,8\,20\,16) \\
19: & (5\,12\,22\,18) & (10\,13\,24\,21) \\
23: & (11\,14\,20\,16) & (18\,21\,24\,22).
\end{array}$$

The four vertices 7, 9, 15 and 17 are not joined to each other and not to any of the pinch points either. Therefore the eight vertex stars of $7, 9, 15, 17, 2, 6, 19, 23$ cover the 64 triangles of the surface entirely and simply, compare Figure 2.5 on the next page where the combinatorial type is sketched. In this drawing all vertices are

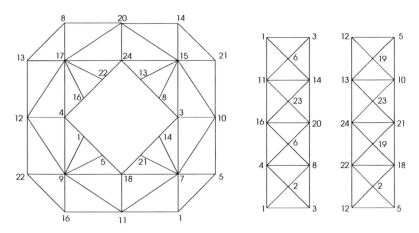

Figure 2.5: The triangulation of the Hamiltonian pinched surface of genus 3 in the
24-cell.

8-valent except for the four pinch points in the two "ladders" on the right hand
side which have to be identified in pairs.

The combinatorial automorphism group of order 16 is generated by

$$Z = (1\,11)(2\,23)(3\,14)(4\,16)(5\,18)(8\,20)(10\,21)(12\,22)(13\,24),$$
$$A = (1\,5)(3\,12)(4\,10)(6\,19)(7\,9)(8\,13)(11\,18)(14\,22)(15\,17)(16\,21)(20\,24),$$
$$B = (1\,3)(4\,8)(5\,10)(9\,15)(11\,14)(12\,13)(16\,20)(18\,21)(22\,24).$$

The elements A and B generate the dihedral group D_8 of order 8 whereas Z
commutes with A and B. Therefore the group is isomorphic with $D_8 \times C_2$.

2.4 Hamiltonian surfaces in the 120-cell and the 600-cell

The 600-cell has the f-vector $(120, 720, 1200, 600)$ and by duality the 120-cell has
the f-vector $(600, 1200, 720, 120)$. Any 1-Hamiltonian surface in the 600-cell must
have 120 vertices, 720 edges and, consequently, 480 triangles (namely, two out of
five), so it has Euler characteristic $\chi = -120$ and genus $g = 61$.

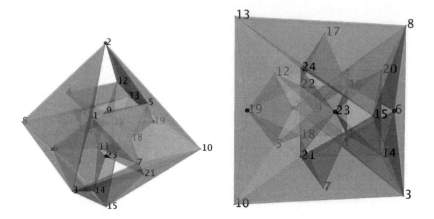

Figure 2.6: Two projections of the Hamiltonian pinched surface of genus 3 in the 24-cell.

We obtain the same genus in the 120-cell by counting 600 vertices, 1200 edges and 480 pentagons (namely, two out of three).

The same Euler characteristic would hold for a pinched surface if there were any. We remark that similarly the 4-cube admits a Hamiltonian surface of the same genus (namely, $g = 1$) as the 4-dimensional cross polytope.

Theorem 2.2

There is no 1-Hamiltonian surface in the 2-skeleton of the 120-cell. There is no pinched surface either since the vertex figure of the 120-cell is too small for containing two disjoint circuits.

The proof is a fairly simple procedure: In each vertex figure of type $\{3,3\}$ (i.e. a tetrahedron) the Hamiltonian surface appears as a Hamiltonian circuit of length 4. This is unique, up to symmetries of the tetrahedron and of the 120-cell itself, see Figure 2.3 on page 46 (right).

Note that two consecutive edges determine the circuit completely. So without loss of generality we can start with such a unique vertex link of the surface. This means we start with four pentagons covering the star of one vertex. In each of the four neighboring vertices this determines two consecutive edges of the link there. It follows that these circuits are uniquely determined as well and that we can extend the beginning part of our surface, now covering the stars of five vertices. Successively this leads to a construction of such a surface. However, after a few steps it ends at a contradiction. Consequently, such a Hamiltonian surface does not exist.

Theorem 2.3

There is no 1-Hamiltonian surface in the 2-skeleton of the 600-cell.

This proof is more involved since it uses the classification of all 17 distinct Hamiltonian circuits in the icosahedron, up to symmetries of it [60, pp. 277 ff.], see Figure 2.7 on the next page.

If there is such a 1-Hamiltonian surface, then the link of each vertex in it must be a Hamiltonian cycle in the vertex figure of the 600-cell which is an icosahedron. We just have to see how these can fit together. Starting with one arbitrary link one can try to extend the triangulation to the neighbors. For the neighbors there are forbidden 2-faces which has a consequence for the possible types among the 17 for them.

After an exhaustive computer search it turned out that there is no way to fit all vertex links together. Therefore such a surface does not exist.

At this point it must be left open whether there are 1-Hamiltonian pinched surfaces in the 600-cell. The reason is that there are too many possibilities for a splitting into two, three or four cycles in the vertex link. For a systematic search one would have to classify all these possibilities first.

The GAP programs used for the algorithmic proofs of Theorems 2.1, 2.2 and 2.3 and details of the calculations are available from the author's website [41] or upon request, see also Appendix D on page 153 for the case of the 24-cell.

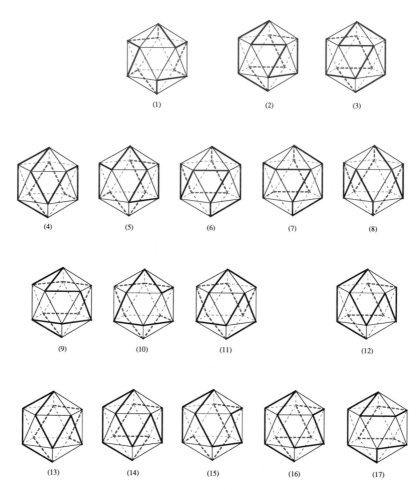

Figure 2.7: The 17 topological types of Hamiltonian cycles in the icosahedron ordered by their symmetries. The cycle (1) has a cyclic symmetry group C_3 of order 3, (2) and (3) have a symmetry of type $C_2 \times C_2$, the cycles (4)-(11) have a C_2 symmetry and (12)-(17) are not symmetric.

Chapter 3

Combinatorial manifolds with stacked vertex links

In even dimensions, super-neighborliness is known to be a purely combinatorial condition which implies the tightness of a triangulation. In this chapter[1] we present other sufficient and purely combinatorial conditions which can be applied to the odd-dimensional case as well. One of the conditions is that all vertex links are stacked spheres, which implies that the triangulation is in Walkup's class $\mathcal{K}(d)$. We show that in any dimension $d \geq 4$ *tight-neighborly* triangulations as defined by Lutz, Sulanke and Swartz are tight. Furthermore, triangulations with k-stacked vertex links are discussed.

In the course of proving the Lower Bound Conjecture (LBC) for 3- and 4-manifolds, D. Walkup [137] defined a class $\mathcal{K}(d)$ of "certain especially simple" [137, p. 1] combinatorial manifolds as the set of all combinatorial d-manifolds that only have *stacked* $(d-1)$-spheres as vertex links as defined below.

Definition 3.1 (stacked polytope, stacked sphere [137])

(i) *A simplex is a* stacked *polytope and each polytope obtained from a stacked polytope by adding a pyramid over one of its facets is again stacked.*

(ii) *A triangulation of the d-sphere S^d is called* stacked d-sphere *if it is combinatorially isomorphic to the boundary complex of a stacked $(d+1)$-polytope.*

---

[1]This chapter essentially contains the results of [42].

Thus, a stacked d-sphere can be understood as the combinatorial manifold obtained from the boundary of the $(d+1)$-simplex by successive stellar subdivisions of facets of the boundary complex $\partial \Delta^{d+1}$ of the $(d+1)$-simplex (i.e. by successively subdividing facets of a complex K_i, $i = 0, 1, 2, \ldots$, by inner vertices, where $K_0 = \partial \Delta^{d+1}$). In the following we will give combinatorial conditions for the tightness of members of $\mathcal{K}(d)$ holding in all dimensions $d \geq 4$. The main results of this chapter are the following.

In Theorem 3.2 on the next page we show that any polar Morse function subject to a condition on the number of critical points of even and odd indices is a perfect function. This can be understood as a combinatorial analogon to Morse's lacunary principle, see Remark 3.3 on page 58.

This result is used in Theorem 3.5 on page 59 in which it is shown that every 2-neighborly member of $\mathcal{K}(d)$ is a tight triangulation for $d \geq 4$. Thus, all *tight-neighborly* triangulations as defined in [93] are tight for $d \geq 4$ (see Section 3.3 on page 62).

This chapter is organized as follows. Section 3.1 on the next page investigates on a certain family of perfect Morse functions. The latter functions can be used to give a combinatorial condition for the tightness of odd-dimensional combinatorial manifolds in terms of properties of the vertex links of such manifolds.

In Section 3.2 on page 58 the tightness of members of $\mathcal{K}(d)$ is discussed, followed by a discussion of the tightness of tight-neighborly triangulations for $d \geq 4$ in Section 3.3 on page 62. Both sections include examples of triangulations for which the stated theorems hold.

In Section 3.4 on page 67 the classes $\mathcal{K}^k(d)$ of combinatorial manifolds are introduced as a generalization of Walkup's class $\mathcal{K}(d)$ and examples of manifolds in these classes are presented. Furthermore, an analogue of Walkup's theorem [137, Thm. 5], [78, Prop. 7.2] for $d = 6$ is proved, assuming the validity of the Generalized Lower Bound Conjecture 3.24 on page 71.

3.1 Polar Morse functions and tightness

There exist quite a few examples of triangulations in even dimensions that are known to be tight (see Section 1.6 on page 26), whereas "for odd-dimensional manifolds it seems to be difficult to transform the tightness of a polyhedral embedding into a simple combinatorial condition", as Kühnel [78, Chap. 5] observed. Consequently there are few examples of triangulations of odd-dimensional manifolds that are known to be tight apart from the sporadic triangulations in [84] and Kühnel's infinite series of $S^{d-1} \times S^1$ for even $d \geq 2$.

It is a well known fact that in even dimensions a Morse function which only has critical points of even indices is a tight function, cf. [25]. This follows directly from the Morse relations, i.e. the fact that $\sum_i (-1)^i \mu_i = \chi(M)$ holds for any Morse function on a manifold M and the fact that $\mu_i \geq \beta_i$. In odd dimensions on the other hand, argumenting in this way is impossible as we always have $\mu_0 \geq 1$ and the alternating sum allows the critical points to cancel out each other. What will be shown in Theorem 3.2 is that at least for a certain family of Morse functions the tightness of its members can readily be determined in arbitrary dimensions $d \geq 3$.

Theorem 3.2

Let $\mathbb{F}$ be any field, $d \geq 3$ and f a polar Morse function on a combinatorial $\mathbb{F}$-orientable d-manifold M such that the number of critical points of f (counted with multiplicity) satisfies

$$\mu_{d-i}(f; \mathbb{F}) = \mu_i(f; \mathbb{F}) = \begin{cases} 0 & \text{for even } 2 \leq i \leq \lfloor \frac{d}{2} \rfloor \\ k_i & \text{for odd } 1 \leq i \leq \lfloor \frac{d}{2} \rfloor \end{cases},$$

where $k_i \geq 0$ for arbitrary d and moreover $k_{\lfloor d/2 \rfloor} = k_{\lceil d/2 \rceil} = 0$, if d is odd. Then f is a tight function.

Proof. Note that as f is polar, M necessarily is connected and orientable. If $d = 3$, $\mu_0 = \mu_3 = 1$ and $\mu_1 = \mu_2 = 0$, and the statement follows immediately. Thus, let us only consider the case $d \geq 4$ from now on. Assume that the vertices $v_1, \ldots, v_n$ of M are ordered by their f-values, $f(v_1) < f(v_2) < \cdots < f(v_n)$. In the long exact sequence for the relative homology

$$\ldots \to H_{i+1}(M_v, M_v\backslash\{v\}) \to H_i(M_v\backslash\{v\}) \overset{\iota_i^*}{\to} H_i(M_v) \to$$
$$\to H_i(M_v, M_v\backslash\{v\}) \to H_{i-1}(M_v\backslash\{v\}) \to \ldots \tag{3.1}$$

the tightness of f is equivalent to the injectivity of the inclusion map ι_i^* for all i and all $v \in V(M)$. The injectivity of ι_i^* means that for any fixed $j = 1, \ldots, n$, the homology $H_i(M_{v_j}, M_{v_{j-1}})$ (where $M_{v_0} = \varnothing$) persists up to the maximal level $H_i(M_{v_n}) = H_i(M)$ and is mapped injectively from level v_j to level v_{j+1}. This obviously is equivalent to the condition for tightness given in Definition 1.42 on page 27. Thus, tight triangulations can also be interpreted as triangulations with the maximal persistence of the homology in all dimensions with respect to the vertex ordering induced by f (see [39]). Hence, showing the tightness of f is equivalent to proving the injectivity of ι_i^* at all vertices $v \in V(M)$ and for all i, what will be done in the following. Note that for all values of i for which $\mu_i = 0$, nothing has to be shown so that we only have to deal with the cases where $\mu_i > 0$ below.

The restriction of the number of critical points being non-zero only in every second dimension results in

$$\dim_{\mathbb{F}} H_i(M_v, M_v\backslash\{v\}) \leq \mu_i(f; \mathbb{F}) = 0$$

and

$$\dim_{\mathbb{F}} H_{d-i}(M_v, M_v\backslash\{v\}) \leq \mu_{d-i}(f; \mathbb{F}) = 0$$

and thus in $H_i(M_v, M_v\backslash\{v\}) = H_{d-i}(M_v, M_v\backslash\{v\}) = 0$ for all even $2 \leq i \leq \lfloor \frac{d}{2} \rfloor$ and all $v \in V(M)$, as M is $\mathbb{F}$-orientable. This implies a splitting of the long exact sequence (3.1) at every second dimension, yielding exact sequences of the forms

$$0 \to H_{i-1}(M_v\backslash\{v\}) \overset{\iota_{i-1}^*}{\to} H_{i-1}(M_v) \to H_{i-1}(M_v, M_v\backslash\{v\}) \to \ldots$$

and

$$0 \;\to\; H_{d-i-1}(M_v\backslash\{v\}) \;\overset{\iota^*_{d-i-1}}{\to}\; H_{d-i-1}(M_v) \;\to\; H_{d-i-1}(M_v, M_v\backslash\{v\}) \;\to\; \ldots,$$

where the inclusions ι^*_{i-1} and ι^*_{d-i-1} are injective for all vertices $v \in V(M)$, again for all even $2 \le i \le \lfloor \frac{d}{2} \rfloor$. Note in particular, that $\mu_{d-2} = 0$ always holds. For critical points of index $d-1$, the situation looks alike:

$$0 \;\to\; \underbrace{H_d(M_v\backslash\{v\})}_{=0} \;\to\; H_d(M_v) \;\to\; H_d(M_v, M_v\backslash\{v\}) \;\to$$

$$\to\; H_{d-1}(M_v\backslash\{v\}) \;\overset{\iota^*_{d-1}}{\to}\; H_{d-1}(M_v) \;\to\; H_{d-1}(M_v, M_v\backslash\{v\}) \;\to\; \ldots$$

By assumption, f only has one maximal vertex as it is polar. Then, if v is not the maximal vertex with respect to f, $H_d(M_v, M_v\backslash\{v\}) = 0$ and thus ι^*_{d-1} is injective. If, on the other hand, v is the maximal vertex with respect to f, one has

$$H_d(M) \cong H_d(M_v, M_v\backslash\{v\}),$$

as $M_v = M$ in this case. Consequently, by the exactness of the sequence above, ι^*_{d-1} is also injective in this case. Altogether it follows that ι^*_i is injective for all i and for all vertices $v \in V(M)$ and thus that f is $\mathbb{F}$-tight. $\qquad\square$

As we will see in Section 3.2 on the following page, this is a condition that can be translated into a purely combinatorial one. Examples of manifolds to which Theorem 3.2 on page 55 applies will be given in the following sections.

3.3 Remark

(*i*) *Theorem 3.2 can be understood as a combinatorial equivalent of Morse's lacunary principle [26, Lecture 2]. The lacunary principle in the smooth case states that if f is a smooth Morse function on a smooth manifold M such that its Morse polynomial $M_t(f)$ contains no consecutive powers of t, then f is a perfect Morse function.*

(*ii*) *Due to the Morse relations, Theorem 3.2 puts a restriction on the topology of manifolds admitting these kinds of Morse functions. In particular, these* *must have vanishing Betti numbers in the dimensions where the number of critical points is zero. Note that in dimension $d = 3$ the theorem thus only holds for homology 3-spheres with $\beta_1 = \beta_2 = 0$ and no statements concerning the tightness of triangulations with $\beta_1 > 0$ can be made. One way of proving the tightness of a 2-neighborly combinatorial 3-manifold M would be to show that the mapping*

$$H_2(M_v) \to H_2(M, M_v \backslash \{v\}) \tag{3.2}$$

is surjective for all $v \in V(M)$ and all rsl functions f. This would result in an injective mapping in the homology group $H_1(M_v \backslash \{v\}) \to H_1(M_v)$ for all $v \in V(M)$ – as above by virtue of the long exact sequence for the relative homology – and thus in the 1-tightness of M, which is equivalent to the ($\mathbb{F}_2$-)tightness of M for $d = 3$, see [78, Prop. 3.18]. Unfortunately, there does not seem to be an easy to check combinatorial condition on M that is sufficient for the surjectivity of the mapping (3.2), in contrast to the case of a combinatorial condition for the 0-tightness of M for which this is just the 2-neighborliness of M.

3.2 Tightness of members of $\mathcal{K}(d)$

In this section we will investigate the tightness of members of Walkup's class $\mathcal{K}(d)$, the family of all combinatorial d-manifolds that only have stacked $(d-1)$-spheres as vertex links. For $d \leq 2$, $\mathcal{K}(d)$ is the set of all triangulated d-manifolds. Kalai [68] showed that the stacking-condition of the links puts a rather strong topological restriction on the members of $\mathcal{K}(d)$:

Theorem 3.4 (Kalai, [68, 11])
Let $d \geq 4$. Then M is a connected member of $\mathcal{K}(d)$ if and only if M is obtained from a stacked d-sphere by $\beta_1(M)$ combinatorial handle additions.

Here a *combinatorial handle addition* to a complex C is defined as usual (see [137, 68, 93]) as the complex C^ψ obtained from C by identifying two facets Δ_1 and Δ_2 of C such that $v \in V(\Delta_1)$ is identified with $w \in \Delta_2$ only if $\mathrm{d}(v, w) \geq 3$, where

$V(X)$ denotes the vertex set of a simplex X and $d(v, w)$ the distance of the vertices v and w in the 1-skeleton of C seen as undirected graph (cf. [9]).

In other words Kalai's theorem states that any connected $M \in \mathcal{K}(d)$ is necessarily homeomorphic to a connected sum with summands of the form $S^1 \times S^{d-1}$ and $S^1 \times S^{d-1}$, compare [93]. Looking at 2-neighborly members of $\mathcal{K}(d)$, the following observation concerning the embedding of the triangulation can be made.

Theorem 3.5

Let $d = 2$ or $d \geq 4$. Then any 2-neighborly member of $\mathcal{K}(d)$ yields a tight triangulation of the underlying PL manifold.

Note that since any triangulated 1-sphere is stacked, $\mathcal{K}(2)$ is the set of all triangulated surfaces and that any 2-neighborly triangulation of a surface is tight. The two conditions of the manifold being 2-neighborly and having only stacked spheres as vertex links are rather strong as the only stacked sphere that is k-neighborly, $k \geq 2$, is the boundary of the simplex, see also Remark 3.20 on page 69. Thus, the only k-neighborly member of $\mathcal{K}(d)$, $k \geq 3$, $d \geq 2$, is the boundary of the $(d + 1)$-simplex.

The following lemma will be needed for the proof of Theorem 3.5.

Lemma 3.6 *Let S be a stacked d-sphere, $d \geq 3$, and $V' \subseteq V(S)$. Then*

$$H_{d-j}(\mathrm{span}_S(V')) = 0 \quad \text{for } 2 \leq j \leq d - 1,$$

where H_* denotes the simplicial homology groups.

Proof. Assume that $S_0 = \partial \Delta^{d+1}$ and assume S_{i+1} to be obtained from S_i by a single stacking operation such that there exists an $N \in \mathbb{N}$ with $S_N = S$. Then S_{i+1} is obtained from S_i by removing a facet of S_i and the boundary of a new d-simplex T_i followed by a gluing operation of S_i and T_i along the boundaries of the removed facets. This process can also be understood in terms of a bistellar 0-move carried out on a facet of S_i. Since this process does not remove any $(d - 1)$-simplices from S_i or T_i we have $\mathrm{skel}_{d-1}(S_i) \subset \mathrm{skel}_{d-1}(S_{i+1})$.

We prove the statement by induction on i. Clearly, the statement is true for $i = 0$, as $S_0 = \partial \Delta^{d+1}$ and $\partial \Delta^{d+1}$ is $(d + 1)$-neighborly. Now assume that the statement holds for S_i and let $V'_{i+1} \subset V(S_{i+1})$. In the following we can consider the connected components C_k of $\operatorname{span}_{S_{i+1}}(V'_{i+1})$ separately. If $C_k \subset S_i$ or $C_k \subset T_i$ then the statement is true by assumption and the $(d+1)$-neighborliness of $\partial\Delta^{d+1}$, respectively. Otherwise let $P_1 := C_k \cap S_i \neq \varnothing$ and $P_2 := C_k \cap T_i \neq \varnothing$. Then

$$H_{d-j}(P_1) \cong H_{d-j}(P_1 \cap T_i) \text{ and } H_{d-j}(P_2) \cong H_{d-j}(P_2 \cap S_i).$$

This yields

$$
\begin{aligned}
H_{d-j}(P_1 \cup P_2) &= H_{d-j}((P_1 \cup P_2) \cap S_i \cap T_i) \\
&= H_{d-j}(\operatorname{span}_{S_i \cap T_i}(V'_{i+1})) \\
&= H_{d-j}(\operatorname{span}_{S_i \cap T_i}(V'_{i+1} \cap V(S_i \cap T_i))) \\
&= 0,
\end{aligned}
$$

as $S_i \cap T_i = \partial\Delta^d$, which is $(d-1)$-neighborly, so that the span of any vertex set has vanishing $(d-j)$-th homology for $2 \leq j \leq d-1$. □

Proof (of Theorem 3.5 on the previous page). For $d = 2$, see [78] for a proof. From now on assume that $d \geq 4$. As can be shown via excision (see, for example [75]), if M is a combinatorial d-manifold, $f : M \to \mathbb{R}$ an rsl function on M and $v \in V(M)$, then

$$H_*(M_v, M_v \backslash \{v\}) \cong H_*(M_v \cap \operatorname{st}(v), M_v \cap \operatorname{lk}(v)).$$

Now let $d \geq 4$, $1 < i < d-1$. The long exact sequence for the relative homology

$$
\begin{aligned}
\cdots &\to H_{d-i}(M_v \cap \operatorname{st}(v)) \to H_{d-i}(M_v \cap \operatorname{st}(v), M_v \cap \operatorname{lk}(v)) \to \\
&\to H_{d-i-1}(M_v \cap \operatorname{lk}(v)) \to H_{d-i-1}(M_v \cap \operatorname{st}(v)) \to \cdots
\end{aligned}
$$

yields an isomorphism

$$H_{d-i}(M_v \cap \operatorname{st}(v), M_v \cap \operatorname{lk}(v)) \cong H_{d-i-1}(M_v \cap \operatorname{lk}(v)), \tag{3.3}$$

as $H_{d-i}(M_v \cap \operatorname{st}(v)) = H_{d-i-1}(M_v \cap \operatorname{st}(v)) = 0$. Note here, that $M_v \cap \operatorname{st}(v)$ is a cone over $M_v \cap \operatorname{lk}(v)$ and thus contractible.

Since $M \in \mathcal{K}(d)$, all vertex links in M are stacked $(d - 1)$-spheres and thus Lemma 3.6 on page 59 applies to the right hand side of (3.3) on the preceding page. This implies that a d-manifold $M \in \mathcal{K}(d)$, $d \geq 4$, cannot have critical points of index $2 \leq i \leq d - 2$, i.e. $\mu_2(f; \mathbb{F}) = \cdots = \mu_{d-2}(f; \mathbb{F}) = 0$.

Furthermore, the 2-neighborliness of M implies that any function on M is polar. Thus, all prerequisites of Theorem 3.2 on page 55 are fulfilled, f is tight and consequently M is a tight triangulation, what was to be shown. □

3.7 Remark *In even dimensions $d \geq 4$, Theorem 3.5 on page 59 can also be proved without using Theorem 3.2 on page 55. In this case the statement follows from the 2-neighborliness of M (that yields $\mu_0(f; \mathbb{F}) = \beta_0$ and $\mu_d(f; \mathbb{F}) = \beta_d$), and the Morse relations 1.57 on page 36 which then yield $\mu_1(f; \mathbb{F}) = \beta_1$ and $\mu_{d-1}(f; \mathbb{F}) = \beta_{d-1}$ for any rsl function f, as $\mu_2(f; \mathbb{F}) = \cdots = \mu_{d-2}(f; \mathbb{F}) = 0$.*

As a consequence, the stacking condition of the links already implies the vanishing of $\beta_2, \ldots, \beta_{d-2}$ (as by the Morse relations $\mu_i \geq \beta_i$), in accordance with Kalai's Theorem 3.4 on page 58.

An example of a series of tight combinatorial manifolds is the infinite series of sphere bundles over the circle due to Kühnel [74]. The triangulations in this series are all 2-neighborly on $f_0 = 2d + 3$ vertices. They are homeomorphic to $S^{d-1} \times S^1$ in even dimensions and to $S^{d-1} \rtimes S^1$ in odd dimensions. Furthermore, all links are stacked and thus Theorem 3.5 applies providing an alternative proof of the tightness of the triangulations in this series.

Corollary 3.8

All members M^d of the series of triangulations in [74] are 2-neighborly and lie in the class $\mathcal{K}(d)$. They are thus tight triangulations by Theorem 3.5.

Another example of a triangulation to which Theorem 3.5 on page 59 applies is due to Bagchi and Datta [11] and will be given in the following section. It is an example of a so called *tight-neighborly* triangulation as defined by Lutz, Sulanke and Swartz [93]. For this class of manifolds, Theorem 3.5 on page 59 holds for $d = 2$ and $d \geq 4$. Tight-neighborly triangulations will be described in more detail in the next section.

3.3 Tight-neighborly triangulations

Beside the class of combinatorial d-manifolds with stacked spheres as vertex links $\mathcal{K}(d)$, Walkup [137] also defined the class $\mathcal{H}(d)$. This is the family of all simplicial complexes that can be obtained from the boundary complex of the $(d+1)$-simplex by a series of zero or more of the following three operations: (i) stellar subdivision of facets, (ii) combinatorial handle additions and (iii) forming connected sums of objects obtained from the first two operations.

The two classes are closely related. Obviously, the relation $\mathcal{H}(d) \subset \mathcal{K}(d)$ holds. Kalai [68] showed the reverse inclusion $\mathcal{K}(d) \subset \mathcal{H}(d)$ for $d \geq 4$.

Note that the condition of the 2-neighborliness of an $M \in \mathcal{K}(d)$ in Theorem 3.5 on page 59 is equivalent to the first Betti number $\beta_1(M)$ being maximal with respect to the vertex number $f_0(M)$ of M (as a 2-neighborly triangulation does not allow any handle additions). Such manifolds are exactly the cases of equality in [108, Th. 5.2]. In their recent work [93], Lutz, Sulanke and Swartz prove the following theorem[2].

Theorem 3.9 (Theorem 5 in [93])
Let $\mathbb{K}$ be any field and let M be a $\mathbb{K}$-orientable triangulated d-manifold with $d \geq 3$. Then

$$f_0(M) \geq \left\lceil \frac{1}{2} \left(2d + 3 + \sqrt{1 + 4(d+1)(d+2)\beta_1(M;\mathbb{K})} \right) \right\rceil. \qquad (3.4)$$

3.10 Remark *As pointed out in [93], for $d = 2$ inequality (3.4) coincides with Heawood's inequality*

$$f_0(M) \geq \left\lceil \frac{1}{2} \left(7 + \sqrt{49 - 24\chi(M)} \right) \right\rceil$$

if one replaces $\beta_1(M;\mathbb{K})$ by $\frac{1}{2}\beta_1(M;\mathbb{K})$ to account for the double counting of the middle Betti number $\beta_1(M;\mathbb{K})$ of surfaces by Poincaré duality. Inequality (3.4) can

---

[2]The author would like to thank Frank Lutz for fruitful discussions about tight-neighborly triangulations and pointing him to the work [93] in the first place.

also be written in the form

$$\binom{f_0 - d - 1}{2} \geq \binom{d + 2}{2}\beta_1.$$

Thus, Theorem 5 in [93] settles Kühnel's conjectured bounds

$$\binom{f_0 - d + j - 2}{j + 1} \geq \binom{d + 2}{j + 1}\beta_j \quad with \quad 1 \leq j \leq \lfloor\frac{d - 1}{2}\rfloor$$

in the case $j = 1$.

For $\beta_1 = 1$, the bound (3.4) on the preceding page coincides with the Brehm-Kühnel bound $f_0 \geq 2d + 4 - j$ for $(j - 1)$-connected but not j-connected d-manifolds in the case $j = 1$, see [27]. Inequality (3.4) is sharp by the series of vertex minimal triangulations of sphere bundles over the circle presented in [74].

Triangulations of connected sums of sphere bundles $(S^2 \times S^1)^{\#k}$ and $(S^2 \rtimes S^1)^{\#k}$ attaining equality in (3.4) on the facing page for $d = 3$ were discussed in [93]. Such triangulations are necessarily 2-neighborly and Lutz, Sulanke and Swartz defined the following.

Definition 3.11 (tight-neighborly triangulation, [93]) *Let $d \geq 2$ and let M be a triangulation of $(S^{d-1} \times S^1)^{\#k}$ or $(S^{d-1} \rtimes S^1)^{\#k}$ attaining equality in (3.4) on the preceding page. Then M is called a tight-neighborly triangulation.*

For $d \geq 4$, all triangulations of $\mathbb{F}$-orientable $\mathbb{F}$-homology d-manifolds with equality in (3.4) lie in $\mathcal{H}(d)$ and are tight-neighborly triangulations of $(S^{d-1} \times S^1)^{\#k}$ or $(S^{d-1} \rtimes S^1)^{\#k}$ by Theorem 5.2 in [108].

The authors conjectured [93, Conj. 13] that all tight-neighborly triangulations are tight in the classical sense of Definition 1.42 on page 27 and showed that the conjecture holds in the following cases: for $\beta_1 = 0, 1$ and any d and for $d = 2$ and any β_1. Indeed, the conjecture also holds for any $d \geq 4$ and any β_1 as it is a direct consequence of Theorem 3.5 on page 59.

Corollary 3.12

For $d \geq 4$ all tight-neighborly triangulations are tight.

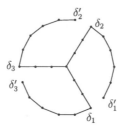

Figure 3.1: Dual graph of the 5-ball B_{30}^5 presented in [11]. In order to obtain M_{15}^4 from the boundary of B_{30}^5, three handles are added over the facet pairs (δ_i, δ_i'), $1 \leq i \leq 3$.

Proof. For $d \geq 4$ one has $\mathcal{H}(d) = \mathcal{K}(d)$ and the statement is true for all 2-neighborly members of $\mathcal{K}(d)$ by Theorem 3.5 on page 59. □

It remains to be investigated whether for vertex minimal triangulations of d-handlebodies, $d \geq 3$, the reverse implication is true, too, i.e. that for this class of triangulations the terms of tightness and tight-neighborliness are equivalent.

Question 3.13 *Let $d \geq 4$ and let M be a tight triangulation homeomorphic to $(S^{d-1} \times S^1)^{\#k}$ or $(S^{d-1} \rtimes S^1)^{\#k}$. Does this imply that M is tight-neighborly?*

As was shown in [93], at least for values of $\beta_1 = 0, 1$ and any d and for $d = 2$ and any β_1 this is true.

One example of a triangulation for which Theorem 3.5 on page 59 holds, is due to Bagchi and Datta [11]. The triangulation M_{15}^4 of $(S^3 \rtimes S^1)^{\#3}$ from [11] is a 2-neighborly combinatorial 4-manifold on 15 vertices that is a member of $\mathcal{K}(4)$ with f-vector $f = (15, 105, 230, 240, 96)$. Since M_{15}^4 is tight-neighborly, we have the following corollary.

Corollary 3.14

The 4-manifold M_{15}^4 given in [11] is a tight triangulation.

Bagchi and Datta constructed M_{15}^4 from the boundary of a 5-ball B_{30}^5 by three simultaneous handle additions, see Figure 3.1.

The next possible triples of values of β_1, d and n for which a 2-neighborly member of $\mathcal{K}(d)$ could exist (compare [93]) are listed in Table 3.1. Apart from the sporadic

Table 3.1: Known and open cases for β_1, d and n of 2-neighborly members of $\mathcal{K}(d)$.

β_1	d	n	top. type	reference
0	any d	$d+1$	S^{d-1}	$\partial \Delta^d$
1	any even $d \geq 2$	$2d+3$	$S^{d-1} \times S^1$	[74] ($d=2$: [102, 35])
1	any odd $d \geq 2$	$2d+3$	$S^{d-1} \rtimes S^1$	[74] ($d=3$: [137, 5])
2	13	35	?	
3	4	15	$(S^3 \rtimes S^1)^{\#3}$	[11]
5	5	21	?	
8	10	44	?	

examples in dimension 4 and the infinite series of higher dimensional analogues of Császár's torus in arbitrary dimension $d \geq 2$ due to Kühnel [74], cf. [83, 9, 33], mentioned earlier, no further examples are known as of today.

Especially in (the odd) dimension $d = 3$, things seem to be a bit more subtle, as already laid out in Remark 3.3 on page 58. As Altshuler and Steinberg [4] showed that the link of any vertex in a neighborly 4-polytope is stacked (compare also Remark 8.5 in [68]), we know that the class $\mathcal{K}(3)$ is rather big compared to $\mathcal{H}(3)$. Thus, a statement equivalent to Theorem 3.5 on page 59 is not surprisingly false for members of $\mathcal{K}(3)$, a counterexample being the boundary of the cyclic polytope $\partial C(4,6) \in \mathcal{K}(3)$ which is 2-neighborly but certainly not a tight triangulation as it has empty triangles. The only currently known example of a tight-neighborly combinatorial 3-manifold is a 9-vertex triangulation M^3 of $S^2 \rtimes S^1$ independently found by Walkup [137] and Altshuler and Steinberg [5]. This triangulation is combinatorially unique, as was shown by Bagchi and Datta [10]. For $d = 3$, it is open whether there exist tight-neighborly triangulations for higher values of $\beta_1 \geq 2$, see [93, Question 12].

The fact that M^3 is a tight triangulation is well known, see [78]. Yet, we will present here another proof of the tightness of M^3. It is a rather easy procedure when looking at the 4-polytope P the boundary of which M^3 was constructed from by one elementary combinatorial handle addition, see also [11].

Lemma 3.15 *Walkup's 9-vertex triangulation M^3 of $S^2 \times S^1$ is tight.*

Proof. Take the stacked 4-polytope P with f-vector $f(P) = (13, 42, 58, 37, 9)$ from [137]. Its facets are

$$\langle 1\,2\,3\,4\,5 \rangle, \qquad \langle 2\,3\,4\,5\,6 \rangle, \qquad \langle 3\,4\,5\,6\,7 \rangle,$$
$$\langle 4\,5\,6\,7\,8 \rangle, \qquad \langle 5\,6\,7\,8\,9 \rangle, \qquad \langle 6\,7\,8\,9\,10 \rangle,$$
$$\langle 7\,8\,9\,10\,11 \rangle, \quad \langle 8\,9\,10\,11\,12 \rangle, \quad \langle 9\,10\,11\,12\,13 \rangle.$$

As P is stacked it has missing edges (called *diagonals*), but no empty faces of higher dimension.

Take the boundary ∂P of P. By construction, P has no inner i-faces, $i \leq 2$ so that ∂P has the 36 diagonals of P and additionally 8 empty tetrahedra, but no empty triangles. As ∂P is a 3-sphere, the empty tetrahedra are all homologous to zero.

Now form a 1-handle over ∂P by removing the two tetrahedra $\langle 1, 2, 3, 4 \rangle$ and $\langle 10, 11, 12, 13 \rangle$ from ∂P followed by an identification of the four vertex pairs $(i, i+9)$, $1 \leq i \leq 4$, where the newly identified vertices are labeled with $1, \ldots, 4$.

This process yields a 2-neighborly combinatorial manifold M^3 with $13-4 = 9$ vertices and one additional empty tetrahedron $\langle 1, 2, 3, 4 \rangle$, which is the generator of $H_2(M)$.

As M^3 is 2-neighborly it is 0-tight and as ∂P had no empty triangles, two empty triangles in the span of any vertex subset $V' \subset V(M)$ are always homologous. Thus, M^3 is a tight triangulation. $\qquad \square$

The construction in the proof above could probably be used in the general case with $d = 3$ and $\beta_1 \geq 2$: one starts with a stacked 3-sphere M_0 as the boundary of a stacked 4-polytope which by construction does not contain empty 2-faces and then successively forms handles over this boundary 3-sphere (obtaining triangulated manifolds $M_1, \ldots, M_n = M$) until the resulting triangulation M is 2-neighborly and fulfills equality in (3.4) on page 62. Note that this can only be done in the regular cases of (3.4), i.e. where (3.4) admits integer solutions for the case of equality. For a list of possible configurations see [93].

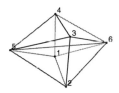

Figure 3.2: A minimally 2-stacked S^2 as the boundary complex of a subdivided 3-octahedron.

3.4 k-stacked spheres and the class $\mathcal{K}^k(d)$

McMullen and Walkup [98] extended the notion of stacked polytopes to k-*stacked polytopes* as simplicial d-polytopes that can be triangulated without introducing new j-faces for $0 \leq j \leq d - k - 1$. More generally, we can define the following.

Definition 3.16 (k-stacked balls and spheres, [98, 68]) *A k-stacked $(d+1)$- ball, $0 \leq k \leq d$, is a triangulated $(d+1)$-ball that has no interior j-faces, $0 \leq j \leq d-k$. A minimally k-stacked $(d+1)$-ball is a k-stacked $(d+1)$-ball that is not $(k-1)$-stacked. The boundary of any (minimally) k-stacked $(d+1)$-ball is called a* (minimally) *k-stacked d-sphere.*

Note that in this context the ordinary stacked d-spheres are exactly the 1-stacked d-spheres. Note also that a k-stacked d-sphere is obviously also $(k+l)$-stacked, where $l \in \mathbb{N}$, $k+l \leq d$, compare [8]. The simplex Δ^{d+1} is the only 0-stacked $(d+1)$-ball and the boundary of the simplex $\partial \Delta^{d+1}$ is the only 0-stacked d-sphere. Keep in mind that all triangulated d-spheres are at least d-stacked [8, Rem. 9.1].

Figure 3.2 shows the boundary of an octahedron as an example of a minimally 2-stacked 2-sphere S with 6 vertices. The octahedron that is subdivided along the inner diagonal $(5,6)$ can be regarded as a triangulated 3-ball B with $\mathrm{skel}_0(S) = \mathrm{skel}_0(B)$ and $\partial B = S$. Note that although all vertices of B are on the boundary, there is an inner edge so that the boundary is 2-stacked, but not 1-stacked. In higher dimensions, examples of minimally d-stacked d-spheres exist as boundary complexes of subdivided d-cross polytopes with an inner diagonal.

Akin to the 1-stacked case, a more geometrical characterization of k-stacked d-spheres can be given via bistellar moves (see Section 1.5 on page 24), at least for $k \leq \lceil \frac{d}{2} \rceil$. Note that for any bistellar move $\Phi_A(M)$, $A * B$ forms a $(d+1)$-simplex. Thus, any sequence of bistellar moves defines a sequence of $(d+1)$-simplices – this we will call the *induced sequence of $(d+1)$-simplices* in the following.

The characterization of k-stacked d-spheres using bistellar moves is the following.

Lemma 3.17 *For $k \leq \lceil \frac{d}{2} \rceil$, a complex S obtained from the boundary of the $(d+1)$-simplex by a sequence of bistellar i-moves, $0 \leq i \leq k-1$, is a k-stacked d-sphere.*

Proof. As $k \leq \lceil \frac{d}{2} \rceil$, the sequence of $(d+1)$-simplices induced by the sequence of bistellar moves is duplicate free and defines a simplicial $(d+1)$-ball B with $\partial B = S$. Furthermore, $\mathrm{skel}_{d-k}(B) = \mathrm{skel}_{d-k}(S)$ holds as no bistellar move in the sequence can contribute an inner j-face to B, $0 \leq j \leq d-k$. Thus, S is a k-stacked d-sphere.□

Keep in mind though, that this interpretation does not hold for values $k > \lceil \frac{d}{2} \rceil$ as in this case the sequence of $(d+1)$-simplices induced by the sequence of bistellar moves may have duplicate entries, as opposed to the case with $k \leq \lceil \frac{d}{2} \rceil$.

In terms of bistellar moves, the minimally 2-stacked sphere in Figure 3.2 on the preceding page can be constructed as follows: Start with a solid tetrahedron and stack another tetrahedron onto one of its facets (a 0-move). Now introduce the inner diagonal $(5,6)$ via a bistellar 1-move. Clearly, this complex is not bistellarly equivalent to the simplex by only applying reverse 0-moves (and thus not (1-)stacked) but it is bistellarly equivalent to the simplex by solely applying reverse 0-, and 1-moves and thus minimally 2-stacked.

The author is one of the authors of the toolkit `simpcomp` [44, 45] for simplicial constructions in the `GAP` system [51]. `simpcomp` contains a randomized algorithm that checks whether a given d-sphere is k-stacked, $k \leq \lceil \frac{d}{2} \rceil$, using the argument above.

With the notion of k-stacked spheres at hand we can define a generalization of Walkup's class $\mathcal{K}(d)$.

Definition 3.18 (the class $\mathcal{K}^k(d)$) *Let $\mathcal{K}^k(d)$, $k \leq d$, be the family of all d-dimensional simplicial complexes all whose vertex links are k-stacked spheres.*

Note that $\mathcal{K}^d(d)$ is the set of all triangulated manifolds for any d and that Walkup's class $\mathcal{K}(d)$ coincides with $\mathcal{K}^1(d)$ above. In analogy to the 1-stacked case, a $(k+1)$-neighborly member of $\mathcal{K}^k(d)$ with $d \geq 2k$ necessarily has vanishing $\beta_1, \ldots, \beta_{k-1}$. Thus, it seems reasonable to ask for the existence of a generalization of Kalai's Theorem 3.4 on page 58 to the class of $\mathcal{K}^k(d)$ for $k \geq 2$.

Furthermore, one might be tempted to ask for a generalization of Theorem 3.5 on page 59 to the class $\mathcal{K}^k(d)$ for $k \geq 2$. Unfortunately, there seems to be no direct way of generalizing Theorem 3.5 to also hold for members of $\mathcal{K}^k(d)$ giving a combinatorial condition for the tightness of such triangulations. The key obstruction here is the fact that a generalization of Lemma 3.6 on page 59 is impossible. While in the case of ordinary stacked spheres a bistellar 0-move does not introduce inner simplices to the $(d-1)$-skeleton, the key argument in Lemma 3.6, this is not true for bistellar i-moves for $i \geq 1$.

Nonetheless, an analogous result to Theorem 3.5 should be true for such triangulations.

Question 3.19 *Let $d \geq 4$ and $2 \leq k \leq \lfloor \frac{d+1}{2} \rfloor$ and let M be a $(k+1)$-neighborly combinatorial manifold such that $M \in \mathcal{K}^k(d)$. Does this imply the tightness of M?*

3.20 Remark *Note that all vertex links of $(k+1)$-neighborly members of $\mathcal{K}^k(d)$ are k-stacked k-neighborly spheres. McMullen and Walkup [98, Sect. 3] showed that there exist k-stacked k-neighborly $(d-1)$-spheres on n vertices for any $2 \leq 2k \leq d < n$. Some examples of such spheres will be given in the following. The conditions of being k-stacked and k-neighborly at the same time is strong as the two conditions tend to exclude each other in the following sense: McMullen and Walkup showed that if a d-sphere is k-stacked and k'-neighborly with $k' > k$, then it is the boundary of the simplex. In that sense the k-stacked k-neighborly spheres appear as the most strongly restricted non-trivial objects of this class: The conditions in Theorem 3.5 on page 59 (with $k = 1$) and in Question 3.19 are the most restrictive ones still admitting non-trivial solutions. If one asks that the links are minimally l-stacked with $l < k$ instead of minimally k-stacked or if one demands the complexes to be $(k+m)$-neighborly, $m > 1$, instead of just $(k+1)$-neighborly, this only leaves the boundary complex of the simplex as a possible solution.*

Table 3.2: Some known tight triangulations and their membership in the classes $\mathcal{K}^k(d)$, cf. [84], with n denoting the number of vertices of the triangulation and $nb.$ its neighborliness.

d	top. type	n	nb.	k
4	$\mathbb{C}P^2$	9	3	2
4	$K3$	16	3	2
4	$(S^3 \times S^1)\#(\mathbb{C}P^2)^{\#5}$	15	2	2
5	$S^3 \times S^2$	12	3	2
5	$SU(3)/SO(3)$	13	3	3
6	$S^3 \times S^3$	13	4	3

Kühnel and Lutz [84] gave an overview of the currently known tight triangulations. The statement of Question 3.19 on the preceding page holds for all the triangulations listed in [84]. Note that there even exist k-neighborly triangulations in $\mathcal{K}^k(d)$ that are tight and thus fail to fulfill the prerequisites of Question 3.19 (see Table 3.2).

Although we did not succeed in proving conditions for the tightness of triangulations lying in $\mathcal{K}^k(d)$, $k \geq 2$, these have nonetheless interesting properties that we will investigate upon in the following. Also, many known tight triangulations are members of these classes, as will be shown. Our first observation is that the neighborliness of a triangulation is closely related to the property of being a member of $\mathcal{K}^k(d)$.

Lemma 3.21 Let $k \in \mathbb{N}$ and M be a combinatorial d-manifold, $d \geq 2k$, that is a $(k+1)$-neighborly triangulation. Then $M \in \mathcal{K}^{d-k}(d)$.

Proof. If M is $(k+1)$-neighborly, then for any $v \in V(M)$, $\mathrm{lk}(v)$ is k-neighborly. As $\mathrm{lk}(v)$ is PL homeomorphic to $\partial\Delta^d$ (since M is a combinatorial manifold) there exists a d-ball B with $\partial B = \mathrm{lk}(v)$ (cf. [8]). Since $\mathrm{lk}(v)$ is k-neighborly, $\mathrm{skel}_{k-1}(B) = \mathrm{skel}_{k-1}(\mathrm{lk}(v))$. By Definition 3.16 on page 67 the link of every vertex $v \in V(M)$ then is $(d-k)$-stacked and thus $M \in \mathcal{K}^{d-k}(d)$. $\qquad\square$

Kühnel [78, Chap. 4] investigated $(k+1)$-neighborly triangulations of $2k$-manifolds and showed that all these are tight triangulations. By Lemma 3.21 on the preceding page all their vertex links are k-stacked spheres and we have the following result.

Corollary 3.22

Let M be a $(k + 1)$-neighborly (tight) triangulation of a $2k$-manifold. Then M lies in $\mathcal{K}^k(2k)$.

In particular this holds for many vertex minimal (tight) triangulations of 4-manifolds.

Corollary 3.23

The known examples of the vertex-minimal tight triangulation of a $K3$-surface with f-vector $f = (16, 120, 560, 720, 288)$ due to Casella and Kühnel [29] and the unique vertex-minimal tight triangulation of $\mathbb{C}P^2$ with f-vector $f = (9, 36, 84, 90, 36)$ due to Kühnel [82], cf. [81] are 3-neighborly triangulations that lie in $\mathcal{K}^2(4)$.

Let us now shed some light on properties of members of $\mathcal{K}^2(6)$. First recall that there exists a *Generalized Lower Bound Conjecture* (GLBC) due to McMullen and Walkup as an extension to the classical Lower Bound Theorem for triangulated spheres as follows.

Conjecture 3.24 (GLBC, cf. [98, 8])

For $d \geq 2k + 1$, the face-vector $(f_0, \ldots, f_d)$ of any triangulated d-sphere S satisfies

$$f_j \geq \begin{cases} \sum_{i=-1}^{k-1}(-1)^{k-i+1}\binom{j-i-1}{j-k}\binom{d-i+1}{j-i}f_i, & \text{if } k \leq j \leq d-k, \\ \sum_{i=-1}^{k-1}(-1)^{k-i+1}\left[\binom{j-i-1}{j-k}\binom{d-i+1}{j-i}\right. \\ \left. -\binom{k}{d-j+1}\binom{d-i}{d-k+1}\right. \\ \left. +\sum_{l=d-j}^{k+1}(-1)^{k-l}\binom{l}{d-j}\binom{d-i}{d-l+1}\right]f_i, & \text{if } d-k+1 \leq j \leq d. \end{cases} \tag{3.5}$$

Equality holds here for any j if and only if S is a k-stacked d-sphere.

The GLBC implies the following theorem for $d = 6$, which is a 6-dimensional analogue of Walkup's theorem [137, Thm. 5], [78, Prop. 7.2], see also Swartz' Theorem 4.10 in [135].

Theorem 3.25

Assuming the validity of the Generalized Lower Bound Conjecture 3.24 on the preceding page, for any combinatorial 6-manifold M the inequality

$$f_2(M) \geq 28\chi(M) - 21f_0 + 6f_1 \tag{3.6}$$

holds. If M is 2-neighborly, then

$$f_2(M) \geq 28\chi(M) + 3f_0(f_0 - 8) \tag{3.7}$$

holds. In either case equality is attained if and only if $M \in \mathcal{K}^2(6)$.

Proof. Clearly,

$$f_3(M) = \frac{1}{4} \sum_{v \in V(M)} f_2(\text{lk}(v)). \tag{3.8}$$

By applying the GLBC 3.24 on the previous page to all the vertex links of M one obtains a lower bound on $f_2(\text{lk}(v))$ for all $v \in V(M)$:

$$f_2(\text{lk}(v)) \geq 35 - 15f_0(\text{lk}(v)) + 5f_1(\text{lk}(v)). \tag{3.9}$$

Here equality is attained if and only if $\text{lk}(v)$ is 2-stacked. Combining (3.8) and (3.9) yields a lower bound

$$\begin{aligned} f_3(M) &\geq \tfrac{1}{4} \sum_{v \in V(M)} 35 - 15f_0(\text{lk}(v)) + 5f_1(\text{lk}(v)) \\ &= \tfrac{5}{4}\left(7f_0(M) - 6f_1(M) + 3f_2(M)\right), \end{aligned} \tag{3.10}$$

for which equality holds if and only if $M \in \mathcal{K}^2(6)$.

If we eliminate f_4, f_5 and f_6 from the Dehn-Sommerville-equations for combinatorial 6-manifolds, we obtain the linear equation

$$35f_0 - 15f_1 + 5f_2 - f_3 = 35\chi(M). \tag{3.11}$$

Inserting inequality (3.10) into (3.11) and solving for $f_2(M)$ yields the claimed lower bounds (3.6) and (3.7) on the preceding page,

$$
\begin{aligned}
f_2(M) &\geq 28\chi(M) - 21 f_0(M) + 6 f_1(M) \\
&= 28\chi(M) + 3 f_0 \underbrace{(f_0(M) - 8)}_{\geq 0},
\end{aligned} \tag{3.12}
$$

where the 2-neighborliness of M was used in the last line. □

For a possible 14-vertex triangulation of $S^4 \times S^2$ (with $\chi = 4$) Inequality (3.12) becomes

$$
f_2 \geq 4 \cdot 28 + 3 \cdot 14 \cdot (14 - 8) = 364,
$$

but together with the trivial upper bound $f_2 \leq \binom{f_0}{3}$ this already would imply that such a triangulation necessarily is 3-neighborly, as $\binom{14}{3} = 364$.

So, just by asking for a 2-neighborly combinatorial $S^4 \times S^2$ on 14 vertices that lies in $\mathcal{K}^2(6)$ already implies that this triangulation is 3-neighborly. Also, the example would attain equality in the Brehm-Kühnel bound [27] as an example of a 1-connected 6-manifold with 14 vertices. We strongly conjecture that this triangulation is tight, see Question 3.19 on page 69.

Chapter 4

Hamiltonian submanifolds of cross polytopes

In this chapter, we investigate Hamiltonian submanifolds of cross polytopes and their properties more closely[1].

The *d-dimensional cross polytope (or d-octahedron)* β^d is defined as the convex hull of the $2d$ points

$$x_i^{\pm} := (0, \ldots, 0, \pm 1, 0, \ldots, 0) \in \mathbb{R}^d.$$

It is a simplicial and regular polytope and it is centrally-symmetric with d missing edges called *diagonals*, each between two antipodal points of type x_i^+ and x_i^-. Its edge graph is the complete d-partite graph with two vertices in each partition, sometimes denoted by $K_2 * \cdots * K_2$. See Figure 4.1 on the following page for an illustration of the cross polytope in dimensions $d = 1, 2, 3$ and Figures 2.1 on page 44 and 2.2 on page 45 for a visualization of the boundary complex of β^4.

The d-cross polytope contains all simplexes not containing one of the d diagonal diagonals and its f-vector satisfies the equality

$$f_i(\beta^d) = 2^{i+1} \binom{d}{i+1}, \quad 0 \le i \le d - 1.$$

---

[1]The first two sections of this chapter are essentially contained in [43], a joint work with Wolfgang Kühnel.

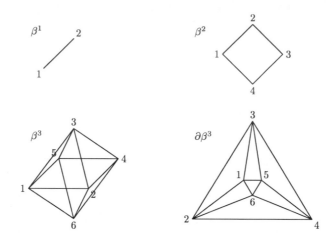

Figure 4.1: The 1-, 2- and 3-cross polytopes β^1, β^2 and β^3 and the boundary complex $\partial\beta^3$ of β^3.

In this chapter, polyhedral manifolds that appear as subcomplexes of the boundary complex of cross polytopes are investigated. Remember that such a subcomplex is called k-Hamiltonian if it contains the full k-skeleton of the polytope. We investigate k-Hamiltonian $2k$-manifolds and in particular 2-Hamiltonian 4-manifolds in the d-dimensional cross polytope. These are the "regular cases" satisfying equality in Sparla's inequality. We present a new example with 16 vertices which is highly symmetric with an automorphism group of order 128. Topologically it is homeomorphic to a connected sum of 7 copies of $S^2 \times S^2$. By this example all regular cases of n vertices with $n < 20$ or, equivalently, all cases of regular d-polytopes with $d \leq 9$ are now decided.

As pointed out in Section 1.6 on page 26, centrally symmetric analogues of tight triangulations appear as Hamiltonian subcomplexes of cross polytopes. A *centrally symmetric triangulation* is a triangulation such that there exists a combinatorial involution operating on the face lattice of the triangulation without fixed points. Any centrally symmetric triangulation thus has an even number of vertices and can be interpreted as a subcomplex of some higher dimensional cross polytope. The tightness of a centrally symmetric $(k-1)$-connected $2k$-manifold M as a subcomplex of β^d then is equivalent to M being a k-Hamiltonian subcomplex of β^d, i.e. to M being *nearly* $(k+1)$-*neighborly*, see [78, Ch. 4].

As it turns out, all of the centrally symmetric triangulations of sphere products $S^l \times S^m$ as k-Hamiltonian subcomplexes of a higher dimensional cross polytope that we investigate in the following lie in the class $\mathcal{K}^{\min\{l,m\}}(d)$, cf. Chapter 3 on page 53. This will be discussed in more detail in Section 4.3 on page 92.

In particular, we present an example of a centrally symmetric triangulation of $S^4 \times S^2 \in \mathcal{K}^2(6)$ as a 2-Hamiltonian subcomplex of the 8-dimensional cross polytope. This triangulation is part of a conjectured series of triangulations of sphere products that are conjectured to be tight subcomplexes of cross polytopes.

4.1 Hamiltonian and tight subcomplexes of cross polytopes

Any 1-Hamiltonian 2-manifold in the d-cross polytope β^d must have the following beginning part of the f-vector:

$$f_0 = 2d, \quad f_1 = 2d(d-1)$$

It follows that the Euler characteristic χ of the 2-manifold satisfies

$$2 - \chi = 2 - 2d + 2d(d-1) - \frac{4}{3}d(d-1) = \frac{2}{3}(d-1)(d-3).$$

These are the regular cases investigated in [66]. In terms of the genus $g = \frac{1}{2}(2-\chi)$ of an orientable surface this equation reads as

$$g = \frac{d-1}{1} \cdot \frac{d-3}{3}.$$

This remains valid for non-orientable surfaces if we assign the genus $\frac{1}{2}$ to the real projective plane. In any case χ can be an integer only if $d \equiv 0, 1 (3)$. The first possibilities, where all cases are actually realized by triangulations of closed orientable surfaces [66], are indicated in Table 4.1 on the following page.

Table 4.1: Regular cases of 1-Hamiltonian 2-manifolds.

d	$2 - \chi$	genus g
3	0	0
4	2	1
6	10	5
7	16	8
9	32	16
10	42	$3 \cdot 7 = 21$
12	66	$3 \cdot 11 = 33$
13	80	$8 \cdot 5 = 40$
15	112	$8 \cdot 7 = 56$

Similarly, any 2-Hamiltonian 4-manifold in the d-cross polytope β^d must have the following beginning part of the f-vector:

$$f_0 = 2d, \quad f_1 = 2d(d-1), f_2 = \frac{4}{3}d(d-1)(d-2)$$

It follows that the Euler characteristic χ satisfies

$$\begin{aligned}
10(\chi - 2) &= f_2 - 4f_1 + 10f_0 - 20 \\
&= \frac{4}{3}d(d-1)(d-2) - 8d(d-1) + 20d - 20 \\
&= \frac{4}{3}(d-1)(d-3)(d-5).
\end{aligned}$$

If we introduce the "genus" $g = \frac{1}{2}(\chi - 2)$ of a simply connected 4-manifold as the number of copies of $S^2 \times S^2$ which are necessary to form a connected sum with Euler characteristic χ, then this equation reads as

$$g = \frac{d-1}{1} \cdot \frac{d-3}{3} \cdot \frac{d-5}{5}.$$

These are the "regular cases". Again the complex projective plane would have genus $\frac{1}{2}$ here. Recall that any 2-Hamiltonian 4-manifold in the boundary of a convex polytope is simply connected since the 2-skeleton is. Therefore the "genus" equals half of the second Betti number.

Moreover, there is an Upper Bound Theorem and a Lower Bound Theorem as follows.

Theorem 4.1 (E. Sparla [125])

If a triangulation of a 4-manifold occurs as a 2-Hamiltonian subcomplex of a centrally-symmetric simplicial d-polytope then the following inequality holds

$$\frac{1}{2}(\chi(M) - 2) \geq \frac{d-1}{1} \cdot \frac{d-3}{3} \cdot \frac{d-5}{5}.$$

Moreover, for $d \geq 6$ equality is possible if and only if the polytope is affinely equivalent to the d-dimensional cross polytope.

If there is a triangulation of a 4-manifold with a fixed point free involution then the number n of vertices is even, i.e., $n = 2d$, and the opposite inequality holds

$$\frac{1}{2}(\chi(M) - 2) \leq \frac{d-1}{1} \cdot \frac{d-3}{3} \cdot \frac{d-5}{5}.$$

Moreover, equality in this inequality implies that the manifold can be regarded as a 2-Hamiltonian subcomplex of the d-dimensional cross polytope.

4.2 Remark *The case of equality in either of these inequalities corresponds to the "regular cases". Sparla's original equation*

$$4^3 \binom{\frac{1}{2}(d-1)}{3} = 10(\chi(M) - 2)$$

is equivalent to the one given above.

By analogy, any k-Hamiltonian $2k$-manifold in the d-dimensional cross polytope satisfies the equation

$$(-1)^k \frac{1}{2}(\chi - 2) = \frac{d-1}{1} \cdot \frac{d-3}{3} \cdot \frac{d-5}{5} \cdot \ldots \cdot \frac{d-2k-1}{2k+1}.$$

It is necessarily $(k-1)$-connected which implies that the left hand side is half of the middle Betti number which is nothing but the "genus".

Furthermore, there is a conjectured Upper Bound Theorem and a Lower Bound Theorem generalizing Theorem 4.1 on the previous page where the inequality has to be replaced by

$$(-1)^k \frac{1}{2}(\chi - 2) \geq \frac{d-1}{1} \cdot \frac{d-3}{3} \cdot \frac{d-5}{5} \cdot \ldots \cdot \frac{d-2k-1}{2k+1}$$

or

$$(-1)^k \frac{1}{2}(\chi - 2) \leq \frac{d-1}{1} \cdot \frac{d-3}{3} \cdot \frac{d-5}{5} \cdot \ldots \cdot \frac{d-2k-1}{2k+1},$$

respectively, see [126], [107].

The discussion of the cases of equality is exactly the same. Sparla's original version

$$4^{k+1} \binom{\frac{1}{2}(d-1)}{k+1} = \binom{2k+1}{k+1}(-1)^k (\chi(M) - 2)$$

is equivalent to the one above. In particular, for any k one of the "regular cases" is the case of a sphere product $S^k \times S^k$ with $(-1)^k(\chi - 2) = 2$ (or "genus" $g = 1$) and $d = 2k + 2$.

So far examples are available for $1 \leq k \leq 4$, even with a vertex transitive automorphism group see [90], [84]. We hope that for $k \geq 5$ there will be similar examples as well, compare Chapter 5 on page 97.

In the case of 2-Hamiltonian subcomplexes of cross polytopes the first non-trivial example was constructed by Sparla as a centrally-symmetric 12-vertex triangulation of $S^2 \times S^2$ as a subcomplex of the boundary of the 6-dimensional cross polytope [125], [88]. Sparla also proved the following analogous Heawood inequality for the case of 2-Hamiltonian 4-manifolds in centrally symmetric d-polytopes

$$\binom{\frac{1}{2}(d-1)}{3} \leq 10(\chi(M) - 2)$$

and the opposite inequality for centrally-symmetric triangulations with $n = 2d$ vertices.

Higher-dimensional examples were found by Lutz [90]: There are centrally-symmetric 16-vertex triangulations of $S^3 \times S^3$ and 20-vertex triangulations of $S^4 \times S^4$. The 2-dimensional example in this series is the well known unique centrally-symmetric 8-vertex torus [79, 3.1]. All these are tightly embedded into the ambient Euclidean space [84].

The generalized Heawood inequality for centrally symmetric $2d$-vertex triangulations of $2k$-manifolds

$$4^{k+1}\binom{\frac{1}{2}(d-1)}{k+1} \geq \binom{2k+1}{k+1}(-1)^k(\chi(M)-2)$$

was conjectured by Sparla in [126] and later almost completely proved by Novik in [107].

Here we show that Sparla's inequality for 2-Hamiltonian 4-manifolds in the skeletons of d-dimensional cross polytopes is sharp for $d \leq 9$. More precisely, we show that each of the regular cases (that is, the cases of equality) for $d \leq 9$ really occurs.

Since the cases $d = 7$ and $d = 9$ are not regular, the crucial point is the existence of an example for $d = 8$ and, necessarily, $\chi = 16$.

Main Theorem 4.3

1. *All cases of* 1*-Hamiltonian surfaces in the regular polytopes are decided. In particular there are no* 1*-Hamiltonian surfaces in the* 24*-cell,* 120*-cell or* 600*-cell.*

2. *All cases of* 2*-Hamiltonian 4-manifolds in the regular d-polytopes are decided up to dimension $d = 9$. In particular, there is a new example of a* 2*-Hamiltonian 4-manifold in the boundary complex of the* 8*-dimensional cross polytope.*

This follows from certain known results and a combination of Theorems 2.1, 2.2, 2.3 in Chapter 2 on page 39, and Theorem 4.4 on the following page.

The regular cases of 1-Hamiltonian surfaces are the following, and each case occurs:

d-simplex: $d \equiv 0, 2\ (3)$ [114]

d-cube: any $d \geq 3$ [19],[112]

d-cross polytope: $d \equiv 0, 1\ (3)$ [66].

The regular cases of 2-Hamiltonian 4-manifolds for $d \leq 9$ are the following:

d-simplex: $d = 5, 8, 9$ [82]

d-cube: $d = 5, 6, 7, 8, 9$ [85]

d-cross polytope: $d = 5, 6, 8$ Theorem 4.4.

Here each of these cases occurs, except for the case of the 9-simplex [82]. Furthermore, 2-Hamiltonian 4-manifolds in the d-cube are known to exist for any $d \geq 5$ [85]. In the case of the d-simplex the next regular case $d = 13$ is undecided, the case $d = 15$ occurs [29]. The next regular case of a d-cross polytope is the case $d = 10$, see Remark 4.6 on page 91.

4.2   2-Hamiltonian 4-manifolds in cross polytopes

In the case of 2-Hamiltonian 4-manifolds as subcomplexes of the d-dimensional cross polytope we have the "regular cases" of equality

$$g = \frac{1}{2}(\chi - 2) = \frac{d-1}{1} \cdot \frac{d-3}{3} \cdot \frac{d-5}{5}.$$

Here χ can be an integer only if $d \equiv 0, 1, 3(5)$. Table 4.2 on the facing page indicates the first possibilities.

Theorem 4.4

There is a 16-vertex triangulation of a 4-manifold $M \cong (S^2 \times S^2)^{\#7}$ which can be regarded as a centrally-symmetric and 2-Hamiltonian subcomplex of the 8-dimensional cross polytope. As one of the "regular cases" it satisfies equality in Sparla's inequalities in Theorem 4.1 on page 79 with the "genus" $g = 7$ and with $d = 8$.

Proof. Any 2-Hamiltonian subcomplex of a convex polytope is simply connected [78, 3.8]. Therefore such an M, if it exists, must be simply connected, in particular $H_1(M) = H_3(M) = 0$.

Table 4.2: Regular cases of 2-Hamiltonian 4-manifolds.

d	$\chi - 2$	"genus" g	existence
5	0	0	$S^4 = \partial \beta^5$
6	2	1	$S^2 \times S^2$ [125],[88]
8	14	7	new, see Theorem 4.4
10	42	$3 \cdot 7 = 21$	see Remark 4.6
11	64	32	?
13	128	64	?
15	224	$16 \cdot 7 = 112$	?
16	286	$11 \cdot 13 = 143$	?
18	442	$13 \cdot 17 = 221$	?
20	646	$17 \cdot 19 = 323$	?
21	720	$8 \cdot 5 \cdot 9 = 360$	?

In accordance with Sparla's inequalities, the Euler characteristic $\chi(M) = 16$ tells us that the middle homology group is $H_2(M, \mathbb{Z}) \cong \mathbb{Z}^{14}$.

The topological type of M is then uniquely determined by the intersection form. If the intersection form is even then by Rokhlin's theorem – which states that the intersection form of any closed PL 4-manifold is divisible by 16 – the signature must be zero, which implies that M is homeomorphic to the connected sum of 7 copies of $S^2 \times S^2$, see [50, 57, 116]. If the intersection form is odd then M is a connected sum of 14 copies of $\pm \mathbb{C}P^2$. We will show that the intersection form of our example is even.

The f-vector $f = (16, 112, 448, 560, 224)$ of this example is uniquely determined already by the requirement of 16 vertices and the condition to be 2-Hamiltonian in the 8-dimensional cross polytope. In particular there are 8 missing edges corresponding to the 8 diagonals of the cross polytope which are pairwise disjoint.

Assuming a vertex-transitive automorphism group, the example was found by using the software of F. H. Lutz described in [90]. The combinatorial automorphism group G of our example is of order 128. With this particular automorphism group the example is unique. The special element

$$\zeta = (1\ 2)(3\ 4)(5\ 6)(7\ 8)(9\ 10)(11\ 12)(13\ 14)(15\ 16)$$

acts on M without fixed points. It interchanges the endpoints of each diagonal and, therefore, can be regarded as the antipodal mapping sending each vertex of the 8-dimensional cross polytope to its antipodal vertex in such a way that it is compatible with the subcomplex M.

A normal subgroup H isomorphic to $C_2 \times C_2 \times C_2 \times C_2$ acts simply transitively on the 16 vertices. The isotropy group G_0 fixing one vertex (and, simultaneously, its antipodal vertex) is isomorphic to the dihedral group of order 8.

The group itself is a semi-direct product between H and G_0. In more detail the example is given by the three G-orbits of the 4-simplices

$$\langle 1\,3\,5\,7\,9\rangle_{128}, \quad \langle 1\,3\,5\,9\,13\rangle_{64}, \quad \langle 1\,3\,5\,7\,15\rangle_{32}$$

with altogether $128 + 64 + 32 = 224$ simplices, each given by a 5-tuple of vertices out of $\{1, 2, 3, \ldots, 15, 16\}$.

The group $G \cong ((((C_4 \times C_2) \rtimes C_2) \rtimes C_2) \rtimes C_2) \rtimes C_2$ of order 128 is generated by the three permutations

$$\alpha = (1\ 12\ 16\ 14\ 2\ 11\ 15\ 13)(3\ 10\ 6\ 8\ 4\ 9\ 5\ 7),$$
$$\beta = (1\ 6\ 2\ 5)(7\ 9\ 2\ 14)(8\ 10\ 11\ 13)(15\ 16),$$
$$\gamma = (1\ 12\ 3\ 14)(2\ 11\ 4\ 13)(5\ 7\ 16\ 10)(6\ 8\ 15\ 9).$$

The complete list of all 224 top-dimensional simplices of M is contained in Section B.1 on page 135.

The link of the vertex 16 is the following simplicial 3 sphere with 70 tetrahedra:

⟨1 3 6 9⟩, ⟨1 3 6 10⟩, ⟨1 3 8 10⟩, ⟨1 3 8 11⟩, ⟨1 3 9 11⟩, ⟨1 4 5 11⟩, ⟨1 4 5 12⟩, ⟨1 4 10 12⟩, ⟨1 4 10 13⟩,⟨1 4 11 13⟩,⟨1 5 10 12⟩,⟨1 5 10 14⟩,⟨1 5 11 14⟩,⟨1 6 7 10⟩, ⟨1 6 7 11⟩, ⟨1 6 9 11⟩, ⟨1 7 10 14⟩,⟨1 7 11 14⟩,⟨1 8 10 13⟩,⟨1 8 11 13⟩,⟨2 3 5 13⟩, ⟨2 3 5 14⟩, ⟨2 3 8 11⟩, ⟨2 3 8 14⟩, ⟨2 3 11 13⟩,⟨2 4 6 7⟩, ⟨2 4 6 8⟩, ⟨2 4 7 13⟩, ⟨2 4 8 10⟩, ⟨2 4 10 13⟩, ⟨2 5 8 12⟩, ⟨2 5 8 14⟩, ⟨2 5 12 13⟩,⟨2 6 7 13⟩, ⟨2 6 8 9⟩, ⟨2 6 9 13⟩, ⟨2 8 9 12⟩, ⟨2 8 10 11⟩, ⟨2 9 12 13⟩,⟨2 10 11 13⟩, ⟨3 5 7 12⟩, ⟨3 5 7 14⟩, ⟨3 5 12 13⟩,⟨3 6 7 10⟩, ⟨3 6 7 12⟩, ⟨3 6 9 12⟩, ⟨3 7 10 14⟩,⟨3 8 10 14⟩, ⟨3 9 11 13⟩,⟨3 9 12 13⟩,⟨4 5 9 12⟩, ⟨4 5 9 14⟩, ⟨4 5 11 14⟩,⟨4 6 7 14⟩, ⟨4 6 8 9⟩, ⟨4 6 9 14⟩, ⟨4 7 11 13⟩,⟨4 7 11 14⟩,⟨4 8 9 12⟩, ⟨4 8 10 12⟩,⟨5 7 9 12⟩, ⟨5 7 9 14⟩, ⟨5 8 10 12⟩,⟨5 8 10 14⟩, ⟨6 7 11 13⟩,⟨6 7 12 14⟩,⟨6 9 11 13⟩,⟨6 9 12 14⟩,⟨7 9 12 14⟩,⟨8 10 11 13⟩.

It remains to prove two facts:

Claim 1. The link of the vertex 16 is a combinatorial 3-sphere. This implies that M is a PL-manifold since all vertices are equivalent under the action of the automorphism group.

A computer algorithm gave a positive answer: the link of the vertex 16 is combinatorially equivalent to the boundary of a 4-simplex by bistellar moves. This method is described in [22] and [90, 1.3].

Claim 2. The intersection form of M is even or, equivalently, the second Stiefel-Whitney class of M vanishes. This implies that M is homeomorphic to the connected sum of 7 copies of $S^2 \times S^2$.

There is an algorithm for calculating the second Stiefel-Whitney class [53]. There are also computer algorithms implemented in **simpcomp** [44, 45] and **polymake** [52], compare [65] for determining the intersection form itself. The latter algorithm gave the following answer: The intersection form of M is even, and the signature is zero. □

In order to illustrate the intersection form on the second homology we consider the link of the vertex 16, as given above. By the tightness condition special homology classes are represented by the empty tetrahedra $c_1 = \langle 7\,10\,11\,16 \rangle$ and $d_1 = \langle 8\,12\,13\,16 \rangle$ which are interchanged by the element

$$\delta = (1\ 2)(5\ 6)(7\ 12)(8\ 11)(9\ 14)(10\ 13)$$

of the automorphism group. The intersection number of these two equals the linking number of the empty triangles $\langle 7\,10\,11 \rangle$ and $\langle 8\,12\,13 \rangle$ in the link of 16. The two subsets in the link spanned by $1, 5, 7, 10, 11, 14$ and $2, 6, 8, 9, 12, 13$, respectively, are homotopy circles interchanged by δ. The intermediate subset of points in the link of 16 which is invariant under δ is the torus depicted in Figure 4.3 on page 89. The set of points which are fixed by δ are represented as the horizontal $(1,1)$-curve in this torus, the element δ itself appears as the reflection along that fixed curve. This torus shrinks down to the homotopy circle on either of the sides which are spanned by $1, 5, 7, 10, 11, 14$ and $2, 6, 8, 9, 12, 13$, respectively.

The empty triangles $\langle 7\,10\,11 \rangle$ and $\langle 8\,12\,13 \rangle$ also represent the same homotopy circles. Since the link is a 3-sphere these two are linked with linking number ± 1. As a result we get for the intersection form $c_1 \cdot d_1 = \pm 1$. These two empty tetrahedra c_1 and d_1 are not homologous to each other in M. Each one can be perturbed into a disjoint position such that the self linking number is zero: $c_1 \cdot c_1 = d_1 \cdot d_1 = 0$. Therefore c_1, d_1 represent a part of the intersection form isomorphic with $\pm \left(\begin{smallmatrix} 0 & 1 \\ 1 & 0 \end{smallmatrix} \right)$.

This situation is transferred to the intersection form of other generators by the automorphism group. As a result we have seven copies of the matrix as a direct sum.

In the homology $H_*(M, \mathbb{Z}) \cong (\mathbb{Z}, 0, \mathbb{Z}^{14}, 0, \mathbb{Z})$ of $(S^2 \times S^2)^{\#7}$ we expect to see 14 generators of $H_2(M)$. In order to visualize M a little bit one can try to visualize the collection of 14 generating homology 2-cycles, even if the intersection form of the manifold cannot be directly derived. These cycles were computed using the computer software `polymake` and are listed in Table 4.3 on the facing page.

One observes that the cycles c_5 and c_7 intersect precisely in the disc D shown in Figure 4.2 on page 88 (top left) and that every other cycle has a non-empty intersection with D, sharing at least one edge with D. Thus, we refer to D as the *universal disc*. In Figure 4.2 on page 88 the cycles c_1 to c_{14} are visualized via their intersection with the universal disc D. In each figure the 1-skeleton of D is shown in form of thin gray lines, the edges shared by D and c_i are shown in green and the edges in the difference $c_i \backslash D$ are shown in blue.

4.5 Remark *Looking at the action of the automorphism group G on the free abelian group $H_2(M, \mathbb{Z}) \cong \mathbb{Z}^{14}$ we get on the 17 conjugacy classes of G the following*

Table 4.3: The 14 homology 2-cycles of M.

cycle c_i	oriented triangles of c_i
c_1	$+\langle 3\,7\,13\rangle$ $-\langle 3\,7\,16\rangle$ $+\langle 3\,10\,12\rangle$ $-\langle 3\,10\,15\rangle$ $+\langle 3\,12\,16\rangle$ $+\langle 3\,13\,15\rangle$ $-\langle 7\,13\,16\rangle$ $-\langle 10\,12\,15\rangle-\langle 12\,13\,15\rangle+\langle 12\,13\,16\rangle$
c_2	$-\langle 3\,12\,14\rangle$ $+\langle 3\,12\,16\rangle$ $-\langle 3\,14\,16\rangle$ $+\langle 8\,12\,14\rangle$ $-\langle 8\,12\,16\rangle$ $+\langle 8\,14\,16\rangle$
c_3	$+\langle 10\,12\,13\rangle-\langle 10\,12\,15\rangle+\langle 10\,13\,15\rangle-\langle 12\,13\,15\rangle$
c_4	$+\langle 5\,9\,11\rangle$ $-\langle 5\,9\,16\rangle$ $+\langle 5\,11\,15\rangle$ $-\langle 5\,13\,15\rangle$ $+\langle 5\,13\,16\rangle$ $-\langle 9\,11\,15\rangle$ $+\langle 9\,13\,15\rangle$ $-\langle 9\,13\,16\rangle$
c_5	$+\langle 4\,8\,14\rangle$ $-\langle 4\,8\,15\rangle$ $+\langle 4\,9\,11\rangle$ $-\langle 4\,9\,16\rangle$ $+\langle 4\,11\,15\rangle$ $+\langle 4\,14\,16\rangle$ $-\langle 8\,12\,15\rangle$ $+\langle 8\,12\,16\rangle$ $-\langle 8\,14\,16\rangle$ $-\langle 9\,11\,15\rangle$ $+\langle 9\,13\,15\rangle$ $-\langle 9\,13\,16\rangle$ $-\langle 12\,13\,15\rangle+\langle 12\,13\,16\rangle$
c_6	$+\langle 6\,10\,12\rangle$ $-\langle 6\,10\,15\rangle$ $+\langle 6\,12\,15\rangle$ $-\langle 10\,12\,15\rangle$
c_7	$-\langle 4\,7\,11\rangle$ $+\langle 4\,7\,16\rangle$ $-\langle 4\,8\,14\rangle$ $+\langle 4\,8\,15\rangle$ $-\langle 4\,11\,15\rangle$ $-\langle 4\,14\,16\rangle$ $+\langle 7\,11\,13\rangle$ $+\langle 7\,13\,16\rangle$ $+\langle 8\,12\,15\rangle$ $-\langle 8\,12\,16\rangle$ $+\langle 8\,14\,16\rangle$ $-\langle 9\,11\,13\rangle$ $+\langle 9\,11\,15\rangle$ $-\langle 9\,13\,15\rangle$ $+\langle 12\,13\,15\rangle-\langle 12\,13\,16\rangle$
c_8	$-\langle 2\,8\,10\rangle$ $+\langle 2\,8\,13\rangle$ $-\langle 2\,10\,15\rangle$ $+\langle 2\,13\,15\rangle$ $+\langle 8\,10\,12\rangle$ $+\langle 8\,12\,15\rangle$ $-\langle 8\,13\,15\rangle$ $-\langle 10\,12\,15\rangle$
c_9	$-\langle 8\,9\,14\rangle$ $+\langle 8\,9\,16\rangle$ $-\langle 8\,14\,16\rangle$ $+\langle 9\,14\,16\rangle$
c_{10}	$-\langle 3\,12\,14\rangle$ $+\langle 3\,12\,16\rangle$ $-\langle 3\,14\,16\rangle$ $-\langle 4\,8\,14\rangle$ $+\langle 4\,8\,15\rangle$ $+\langle 4\,10\,14\rangle$ $-\langle 4\,10\,15\rangle$ $+\langle 8\,12\,15\rangle$ $-\langle 8\,12\,16\rangle$ $+\langle 8\,14\,16\rangle$ $+\langle 10\,12\,14\rangle-\langle 10\,12\,15\rangle$
c_{11}	$-\langle 7\,9\,13\rangle$ $+\langle 7\,9\,16\rangle$ $-\langle 7\,13\,16\rangle$ $+\langle 9\,13\,16\rangle$
c_{12}	$-\langle 9\,11\,13\rangle$ $+\langle 9\,11\,15\rangle$ $-\langle 9\,13\,15\rangle$ $+\langle 11\,13\,15\rangle$
c_{13}	$-\langle 2\,8\,10\rangle$ $+\langle 2\,8\,13\rangle$ $-\langle 2\,10\,15\rangle$ $+\langle 2\,13\,15\rangle$ $+\langle 8\,10\,15\rangle$ $-\langle 8\,13\,15\rangle$
c_{14}	$-\langle 3\,12\,14\rangle$ $+\langle 3\,12\,16\rangle$ $-\langle 3\,14\,16\rangle$ $+\langle 12\,14\,16\rangle$

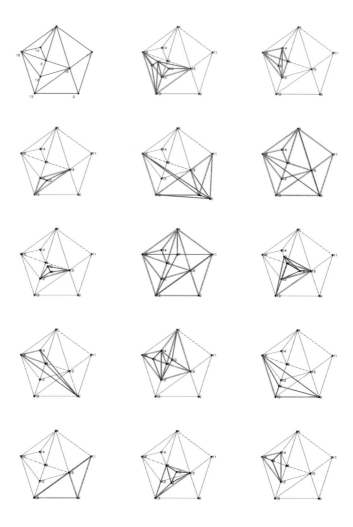

Figure 4.2: Visualization of the 14 homology 2-cycles $c_1, \ldots, c_{14}$ of M (from top to bottom, left to right) via the universal disc D shown in the top left.

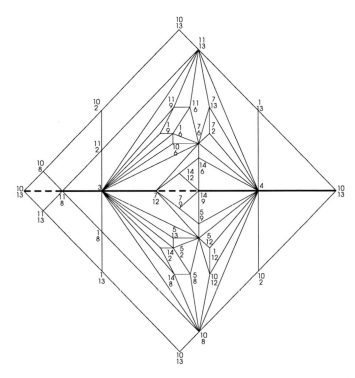

Figure 4.3: The intermediate torus in the link of the vertex 16, invariant under the reflection δ.

character values

$$(14, -2, -2, -2, 2, -2, 6, -2, -2, -2, 6, 0, 0, 0, 0, 0, 0).$$

Denote by χ the corresponding ordinary character. Using the character table[2] of G given by GAP [51] and the orthogonality relations this character decomposes into a sum of five irreducible ordinary characters as follows

$$\chi = \chi_2 + \chi_3 + \chi_{13} + \chi_{14} + \chi_{17}$$

This shows that $\mathbb{C} \otimes_\mathbb{Z} H_2(M, \mathbb{Z})$ is a cyclic $\mathbb{C}G$ - module. It may be interesting to find a geometric explanation for this. The involved irreducible characters are as follows:

	1a	2a	2b	2c	4a	2d	2e	4b	4c	4d	2f	4e	4f	4g	4h	8a	2g
χ_2	1	−1	1	−1	1	1	1	−1	1	−1	1	−1	1	−1	1	−1	1
χ_3	1	−1	1	−1	1	1	1	−1	1	−1	1	1	−1	1	−1	1	−1
χ_{13}	2	.	−2	.	.	2	2	.	−2	.	2	−2	.	2	.	.	.
χ_{14}	2	.	−2	.	.	2	2	.	−2	.	2	2	.	−2	.	.	.
χ_{17}	8	.	.	.	.	−8	.	.	.	.	.	.	.	.	.	.	.

4.6 Remark *There is a real chance to solve the next regular case $d = 10$ in Sparla's inequality. The question is whether there is a 2-Hamiltonian 4-manifold of genus 21 (i.e. $\chi = 44$) in the 10-dimensional cross polytope.*

A 22-vertex triangulation of a manifold with exactly the same genus as a subcomplex of the 11-dimensional cross polytope does exist. If one could save two antipodal vertices by successive bistellar flips one would have a solution.

The example with 22 vertices is defined by the orbits (of length 110 and 22, respectively) of the 4-simplices

$$\langle 1\,3\,5\,7\,18 \rangle_{110}, \quad \langle 1\,3\,5\,7\,21 \rangle_{110}, \quad \langle 1\,3\,5\,8\,18 \rangle_{110},$$
$$\langle 1\,3\,5\,8\,21 \rangle_{110}, \quad \langle 1\,3\,7\,18\,20 \rangle_{110}, \quad \langle 1\,3\,6\,10\,15 \rangle_{22}$$

---

[2]We would like to thank Wolfgang Kimmerle for helpful comments concerning group representations.

under the permutation group of order 110 which is generated by

$$(1 \; 16 \; 7 \; 22 \; 13 \; 5 \; 19 \; 12 \; 3 \; 18 \; 10 \; 2 \; 15 \; 8 \; 21 \; 14 \; 6 \; 20 \; 11 \; 4 \; 17 \; 9)$$

and

$$(1 \; 11 \; 17 \; 3 \; 21)(2 \; 12 \; 18 \; 4 \; 22)(5 \; 9 \; 8 \; 20 \; 14)(6 \; 10 \; 7 \; 19 \; 13).$$

The central involution is

$$(1 \; 2)(3 \; 4)(5 \; 6)(7 \; 8)(9 \; 10)(11 \; 12)(13 \; 14)(15 \; 16)(17 \; 18)(19 \; 20)(21 \; 22)$$

which corresponds to the antipodal mapping in a suitably labeled cross polytope. The f-vector of the example is $(22, 220, 1100, 1430, 572)$, and the middle homology is 42-dimensional, the first and third homology both vanish. Hence it has "genus" 21 in the sense defined above.

Corollary 4.7

There is a tight and PL-taut simplicial embedding of the connected sum of 7 copies of $S^2 \times S^2$ into Euclidean 8-space.

This follows directly from Theorem 4.4 on page 82: The induced polyhedral embedding $M \subset \beta^8 \subset E^8$ into E^8 via the 8-dimensional cross polytope is tight since the intersection with any open half-space is connected and simply connected. No smooth tight embedding of this manifold into 8-space can exist, see [136]. Consequently, this embedding of M into 8-space is smoothable as far as the PL structure is concerned but it is not tightly smoothable.

In addition this example is centrally-symmetric. There is a standard construction of tight embeddings of connected sums of copies of $S^2 \times S^2$ but this works in codimension 2 only, polyhedrally as well as smoothly, see [14, p.101].

The cubical examples in [85] exist in arbitrary codimension but they require a much larger "genus": For a 2-Hamiltonian 4-manifold in the 8-dimensional cube one needs an Euler characteristic $\chi = 64$ which corresponds to a connected sum of 31 copies of $S^2 \times S^2$.

The number of summands in this case grows exponentially with the dimension of the cube. For a 2-Hamiltonian 4-manifold in the 8-dimensional simplex an Euler characteristic $\chi = 3$ is sufficient. It is realized by the 9-vertex triangulation of $\mathbb{C}P^2$ [81], [82]. One copy of $S^2 \times S^2$ cannot be a subcomplex of the 9-dimensional simplex because such a 3-neighborly 10-vertex triangulation does not exist [82] even though it is one of the "regular cases" in the sense of the Heawood type integer condition in Section 1.6 on page 26. In general the idea behind is the following: A given d-dimensional polytope requires a certain minimum "genus" of a $2k$-manifold to cover the full k-dimensional skeleton of the polytope. For the standard polytopes like simplex, d-cube and d-cross polytope we have formulas for the "genus" which is to be expected but we don't yet have examples in all of the cases.

The situation is similar with respect to the concept of tightness: For any given dimension d of an ambient space a certain "genus" of a manifold is required for admitting a tight and substantial embedding into d-dimensional space. This is well understood in the case of 2-dimensional surfaces [78]. For "most" of the simply connected 4-manifolds a tight polyhedral embedding was constructed in [80], without any especially intended restriction concerning the essential codimension. The optimal bounds in this case and in all the other higher-dimensional cases still have to be investigated.

4.3 Subcomplexes with stacked vertex links

In this section we will investigate the relation of the cross polytope and its Hamiltonian submanifolds to the class $\mathcal{K}^k(d)$ of Section 3.4 on page 67.

The boundary of the $(d + 1)$-cross polytope β^{d+1} is an obviously minimally d-stacked d-sphere as it can be obtained as the boundary of a minimally d-stacked $(d + 1)$-ball that is given by any subdivision of β^{d+1} along an inner diagonal.

Corollary 4.8

The 16-vertex triangulation of $(S^2 \times S^2)^{\#7}_{16}$ presented in [43] lies in $\mathcal{K}^2(4)$ and admits a tight embedding into β^8 as shown in [43].

Proof. The triangulation $(S^2 \times S^2)_{16}^{\#7}$ is a combinatorial manifold and a tight subcomplex of β^8 as shown in [43]. Thus, each vertex link is a PL 3-sphere. It remains to show that all vertex links are 2-stacked.

Using the software system simpcomp [44, 45], we found that the vertex links can be obtained from the boundary of a 4-simplex by a sequence of 0- and 1-moves. Therefore, by Lemma 3.17 on page 68, the vertex links are 2-stacked 3-spheres. Thus, $(S^2 \times S^2)_{16}^{\#7} \in \mathcal{K}^2(4)$, as claimed. □

The following centrally symmetric triangulation of $S^4 \times S^2$ is a new example of a triangulation that can be seen as a subcomplex of a higher dimensional cross polytope.

Theorem 4.9

There exists an example of a centrally symmetric triangulation M_{16}^6 of $S^4 \times S^2$ with 16 vertices that is a 2-Hamiltonian subcomplex of the 8-cross polytope β^8 and that lies in $\mathcal{K}^2(6)$.

Proof. The construction of M_{16}^6 was done entirely with simpcomp [44, 45] and is as follows. First a 24-vertex triangulation $\tilde{M}^6$ of $S^4 \times S^2$ was constructed as the standard simplicial cartesian product of $\partial\Delta^3$ and $\partial\Delta^5$ as implemented in [44], where Δ^d denotes the d-simplex. Then $\tilde{M}^6$ obviously is a combinatorial 6-manifold homeomorphic to $S^4 \times S^2$.

This triangulation $\tilde{M}^6$ was then reduced to the triangulation M_{16}^6 with f-vector $f = (16, 112, 448, 980, 1232, 840, 240)$ using a vertex reduction algorithm based on bistellar flips that is implemented in [44]. The code is based on the vertex reduction methods developed by Björner and Lutz [22]. It is well-known that this reduction process leaves the PL type of the triangulation invariant so that $M_{16}^6 \cong S^4 \times S^2$. The 240 5-simplices of M_{16}^6 are given in Section B.2 on page 136. The f-vector of M_{16}^6 is uniquely determined already by the condition of M_{16}^6 to be 2-Hamiltonian in the 8-dimensional cross polytope. In particular, M_{16}^6 has 8 missing edges of the form $\langle i, i+1 \rangle$ for all odd $1 \le i \le 15$, which are pairwise disjoint and correspond to the 8 diagonals of the cross polytope. As there is an involution

$$I = (1,2)(3,4)(5,6)(7,8)(9,10)(11,12)(13,14)(15,16)$$

operating on the faces of M_{16}^6 without fixed points, M^2 can be seen as a 2-Hamiltonian subcomplex of β^8.

It remains to show that $M_{16}^6 \in \mathcal{K}^2(6)$. Remember that the necessary and sufficient condition for a triangulation X to lie in $\mathcal{K}^k(d)$ is that all vertex links of X are k-stacked $(d-1)$-spheres. Since M_{16}^6 is a combinatorial 6-manifold, all vertex links are triangulated 5-spheres. It thus suffices to show that all vertex links are 2-stacked. Using `simpcomp`, we found that the vertex links can be obtained from the boundary of the 6-simplex by a sequence of 0- and 1-moves. Therefore, by Lemma 3.17 on page 68, vertex links are 2-stacked 5-spheres. Thus, $M_{16}^6 \in \mathcal{K}^2(6)$, as

The triangulation M_{16}^6 is strongly conjectured to be tight in β^8. It is part of a conjectured series of centrally symmetric triangulations of sphere products as Hamiltonian subcomplexes of the cross polytope that can be tightly embedded into the cross polytope (see [126], [84, 6.2] and [43, Sect. 6]). In particular the sphere products presented in [84, Thm. 6.3] are part of this conjectured series and the following theorem holds.

Theorem 4.10

The centrally symmetric triangulations of sphere products of the form $S^k \times S^m$ with vertex transitive automorphism group

$$S^1 \times S^1, \quad S^2 \times S^1, \quad S^3 \times S^1, \quad S^4 \times S^1, \quad S^5 \times S^1, \quad S^6 \times S^1, \quad S^7 \times S^1,$$
$$S^2 \times S^2, \quad S^3 \times S^2, \qquad\qquad\quad S^5 \times S^2,$$
$$S^3 \times S^3, \quad S^4 \times S^3, \quad S^5 \times S^3,$$
$$S^4 \times S^4$$

on $n = 2(k+m) + 4$ vertices presented in [84, Theorem 6.3] are all contained in the class $\mathcal{K}^{\min\{k,m\}}(k+m)$.

Using `simpcomp`, we found that the vertex links of all the manifolds mentioned in the statement can be obtained from the boundary of a $(k+m)$-simplex by sequences of bistellar i-moves, $0 \le i \le \min\{k,l\} - 1$. Therefore, by Lemma 3.17 on page 68, the vertex links are $\min\{k,m\}$-stacked $(k+m-1)$-spheres. Thus all the manifolds mentioned in the statement are in $\mathcal{K}^{\min\{k,m\}}(k+m)$. Note that since these examples all have a transitive automorphism group, it suffices to check the stackedness condition for one vertex link only.

The preceding observations naturally lead to the following Question 4.11 as a generalization of Question 3.19 on page 69. Remember that a combinatorial manifold that is $(k + 1)$-neighborly (see Question 3.19) is a k-Hamiltonian subcomplex of a higher dimensional simplex. The following seems to hold for Hamiltonian subcomplexes of cross polytopes in general.

Question 4.11 *Let $d \geq 4$ and let M be a k-Hamiltonian codimension 2 subcomplex of the $(d + 2)$-dimensional cross polytope β^{d+2}, such that $M \in \mathcal{K}^k(d)$ for some fixed $1 \leq k \leq \lceil \frac{d-1}{2} \rceil$. Does this imply that the embedding $M \subset \beta^{d+2} \subset E^{d+2}$ is tight?*

This is true for all currently known codimension 2 subcomplexes of cross polytopes that fulfill the prerequisites of Question 4.11: The 8-vertex triangulation of the torus, a 12-vertex triangulation of $S^2 \times S^2$ due to Sparla [88, 124] and the triangulations of $S^k \times S^k$ on $4k + 4$ vertices for $k = 3$ and $k = 4$ as well as for the infinite series of triangulations of $S^k \times S^1$ in [74]. For the other triangulations of $S^k \times S^m$ listed in Theorem 4.10 on the preceding page, Kühnel and Lutz "strongly conjecture" [84, Sec. 6] that they are tight in the $(k+m+2)$-dimensional cross polytope. Nevertheless it is currently not clear whether the conditions of Question 4.11 imply the tightness of the embedding into the cross polytope.

In accordance with [84, Conjecture 6.2] we then have the following

Conjecture 4.12

Any centrally symmetric combinatorial triangulation of $S^k \times S^m$ on $n = 2(k + m + 2)$ vertices is tight if regarded as a subcomplex of the $\frac{n}{2}$-dimensional cross polytope. The triangulation is contained in the class $\mathcal{K}^{\min\{k,m\}}(k + m)$.

Chapter 5

Centrally symmetric triangulations of sphere products

As far as the integer conditions of the "regular cases" in the generalized Heawood inequalities are concerned, it seems to be plausible to ask for centrally-symmetric triangulations of any sphere product $S^k \times S^l$ with a minimum number of

$$n = 2(k + l + 2)$$

vertices, see Section 4.3 on page 92. In this case each instance can be regarded as a codimension-1-subcomplex of the boundary complex of the $(k + l + 2)$-dimensional cross polytope, and it can be expected to be m-Hamiltonian with $m = \min(k, l)$. This can be understood as a kind of simplicial Hopf decomposition of the $(k+l+1)$-sphere by "Clifford-tori" of type $S^k \times S^l$.

For vertex numbers $n \leq 20$ (i.e., for $k + l \leq 8$) a census of such triangulations with a vertex-transitive automorphism group can be found in [90], compare [84] and Theorem 4.10 on page 94. Here all cases occur except for $S^4 \times S^2$ and $S^6 \times S^2$, and all examples admit a dihedral group action of order $2n$. There exists a triangulation of $S^4 \times S^2$ with non-transitive automorphism group in this series, see Theorem 4.9 on page 93, whereas the case of $S^6 \times S^2$ remains to be found. An infinite series of examples so far seems to be known only for $l = 1$ and arbitrary k. This series due to Kühnel is presented in Section 5.2 on page 102 of this chapter.

97

In Section 5.3 on page 104 we will present a construction principle that is conjectured to yield a series of centrally symmetric triangulations of $S^{k-1} \times S^{k-1}$ as $(k-1)$-Hamiltonian submanifolds of β^{2k}. Before coming to the description of the construction principle, important concepts needed for the construction are discussed, namely cyclic automorphism groups and difference cycles.

5.1 Cyclic automorphism groups and difference cycles

Cyclic automorphism groups play an important role for many combinatorial structures. In this setting, the elements (or in our case: vertices) of a combinatorial object are regarded as elements of $\mathbb{Z}_n$ for some $n \in \mathbb{N}$ and the combinatorial structure consists of a set of tuples over $\mathbb{Z}_n$ which is invariant under the $\mathbb{Z}_n$-action $x \mapsto x + 1$ mod n.

Such structures appear for example in the form of cyclic block designs or cyclic Steiner triple systems in the theory of combinatorial designs, see [20]. Triangulated surfaces with cyclic automorphism group played a crucial role in the proof of the Heawood map color theorem [59, 114], see Section 1.6 on page 26. In the field of polytope theory, cyclic polytopes (which have component-wise maximal f-vector among all polytopes of the same dimension and vertex number) with a cyclic symmetry group appear in the proof of the Upper Bound Theorem, see Chapter 1 on page 1.

If the vertices of a combinatorial manifold M on n vertices are identified with elements of $\mathbb{Z}_n$, then —up to the $\mathbb{Z}_n$ action $x \mapsto x + 1$ mod n— an edge $\langle v_0\, v_1 \rangle$ of M can be encoded by the tuple of differences $(v_1 - v_0, n - v_1 - v_0) = (d, n - d)$, where $d \in \mathbb{Z}_n$ is a non-zero element. Likewise, any k-simplex $\langle v_0 \ldots v_k \rangle$ of M with $v_0 < v_1 < \cdots < v_k$ can be encoded by the tuple of differences $(d_1, \ldots, d_k)$ with non-zero elements $d_i \in \mathbb{Z}_n$.

Definition 5.1 (difference sequences and cycles, cf. [20]) *Let $B \subset \mathbb{Z}_n$ with $B = \{b_1, \ldots, b_k\}$ and assume the representatives to be chosen such that*

$$0 \le b_1 < \cdots < b_k < n.$$

Then the k-difference sequence δB of B is the k-tuple

$$\delta B := (d_1, \dots, d_k) = (b_2 - b_1, \dots, b_k - b_{k-1}, n + b_1 - b_k)$$

and the k-difference cycle ∂B of B is the equivalence class of δB under cyclic permutations. We denote ∂B by

$$(d_1 : \cdots : d_k) = (d_2 : \cdots : d_k : d_1) = \cdots = (d_k : d_1 : \cdots : d_{k-1}).$$

Furthermore, given a difference sequence δB, we will refer to the set of simplices

$$\left\{ \left(x \; x+d_1 \; \dots \; x + \sum_{i=1}^{k} d_i \right) : x \in \mathbb{Z}_n \right\}$$

as realization *of δB. Two difference sequences are called* equivalent *if their realizations are equal as sets. Two difference cycles are called* equivalent, *if there exist two representatives that are equivalent.*

In the following we will drop the prefix k most of the time, just speaking of difference sequences and difference cycles, when actually meaning k-difference sequences and k-difference cycles.

Obviously, for any difference sequence δB we have $0 < d_i < n$ and $\sum_{i=1}^{k} d_i = n$. Note also that, given a k-tuple $(d_1, \dots, d_k)$ that satisfies the properties of Definition 5.1, the set $B := \{0, d_1, d_1 + d_2, \dots, d_1 + \cdots + d_{k-1}\}$ satisfies $\delta B = (d_1, \dots, d_k)$. For any k-element subset $B \subset \mathbb{Z}_n$ and for any $x \in \mathbb{Z}_n$ we have $\partial B = \partial(B + x)$.

Kühnel and Gunter Lassmann [83] extended the notion of a difference cycle of order k to a *permuted difference k-cycle* as a difference k-cycle $(d_{\sigma(1)} : \cdots : d_{\sigma(k)})$ for a given permutation $\sigma \in S_k$ and furthermore defined a *k-permcycle* as the set of permuted difference k-cycles where σ ranges over all possible permutations in the symmetric group S_k.

We can define a natural multiplication on the set of difference cycles that is induced by the multiplication in $\mathbb{Z}_n$. Note that we assume a more general definition of the term difference sequence and cycle below, namely relaxing the prerequisite that $\sum_{i=1}^{k} d_i = n$ for $\delta B = (d_1, \dots, d_k)$ to $\sum_{i=1}^{k} d_i \equiv 0 \mod n$. We will refer to such difference sequences and cycles as *generalized difference sequences* and *generalized difference cycles*, respectively.

Definition 5.2 (multiplication for generalized difference cycles) *Let ∂B be a difference cycle and δB a representative of ∂B. Then a multiplication $\cdot$ can be defined via the mapping*

$$
\cdot: \quad \mathbb{Z}_n{}^\times \quad \times \quad (\mathbb{Z}_n\backslash\{\overline{0}\})^k \quad \rightarrow \quad (\mathbb{Z}_n\backslash\{\overline{0}\})^k
$$
$$
(\overline{x} \quad , \quad (\overline{d_1},\ldots,\overline{d_k})) \quad \mapsto \quad (\overline{xd_1},\ldots,\overline{xd_k}) \quad .
$$

This mapping induces a multiplication on the set of generalized difference cycles. An element $m \in \mathbb{Z}_n$ for which $\partial(m \cdot \delta B) = \partial B$ usually is referred to as multiplier of ∂B *(and $\overline{1} \in \mathbb{Z}_n$ obviously is a multiplier for any ∂B).*

We now show that the multiplication of Definition 5.2 is well-defined even for ordinary difference cycles. Note that for a difference cycle C with $\sum c_i = n$ its multiple $D = xC$ obviously satisfies $\sum d_i \equiv 0 \mod n$, but $\sum d_i = n$ does not necessarily hold.

Definition 5.3 (minimal difference sequences and cycles) *Given any generalized difference sequence $\delta B = (b_1,\ldots,b_k)$ over $\mathbb{Z}_n$, there exists a difference sequence $\delta \hat{B} = (\hat{b}_1,\ldots,\hat{b}_k)$ that is equivalent to δB and for which $\sum_{i=1}^{k} \hat{b}_i = n$ holds. $\delta \hat{B}$ is called the* minimal representation *of δB. Analogously, for any given difference cycle ∂B there exists a* minimal representation $\partial \hat{B}$.

Using the following algorithm, a minimal representation for an arbitrary difference sequence or cycle can be obtained.

Lemma 5.4 (minimal representation of difference cycles) *Let $\delta B = (b_1,\ldots, b_k)$ be a difference sequence over $\mathbb{Z}_n$ with $\sum_{i=1}^{k} b_i \equiv 0 \mod n$. Then the minimal representation of δB can be obtained with the following iterative construction.*

   (i) *If $\sum_{i=1}^{k} b_i = n$, δB is a minimal representation. The iteration stops here. Otherwise continue to* (ii).

   (ii) *Determine a minimal j such that $\sum_{i=1}^{j} b_i > n$ and replace b_j by $\tilde{b}_j := \sum_{i=1}^{j} b_i \mod n - \sum_{i=1}^{j-1} b_i$.*

(iii) If δB contains negative entries, then substitute any triple of successive values $a, -b, c$ with $b > 0$ in B with the triple $a - b, b, c - b$ and iterate this process until there are no more negative entries in δB. Then continue with (i).

The process terminates, as in every iteration at least two entries of the cycle are reduced in magnitude.

The algorithm presented above obviously constructs a minimal representation of any difference cycle after a finite number of steps. Let us give an example of how to obtain a minimal representation of a difference cycle with the construction of Lemma 5.4.

Take $n = 12$ and $\delta B = (1, 1, 1, 1, 8)$, $x = 5$. Then $\delta \tilde{B} = x\delta B = (5, 5, 5, 5, 4)$. $\tilde{B}$ is obviously not a minimal representation, as $\sum_{i=1}^{5} \tilde{b}_i = 24 > 12$. We now apply the algorithm of Lemma 5.4. In the first iteration in step (ii), $j = 3$. We thus get a new difference sequence $(5, 5, -7, 5, 4)$ which in step (iii) is first replaced by the difference sequence $(5, -2, 7, -2, 4)$, which in turn is replaced by $(3, 2, 3, 2, 2)$. This last difference sequence $\delta \hat{B} = (3, 2, 3, 2, 2)$ is a minimal representation of δB, as $\sum_{i=1}^{5} \hat{b}_i = 12$.

We will assume that a difference sequence or cycle is given in its minimal representation from now on unless stated otherwise.

An easy calculation (cf. [128, Prop. 1.5]) lets us prove the following lemma.

Lemma 5.5 (length of difference cycles, [128]) *Let $\partial D = (d_1 : \cdots : d_k)$ be a k-difference cycle over $\mathbb{Z}_n$ and let $1 \leq j \leq k$ be the smallest integer such that $j|k$ and $d_i = d_{i+j}$ for all $1 \leq i \leq k - j$. Then the length of the realization of any representative δD of ∂D is*

$$L(\partial D) := \sum_{i=1}^{j} d_i = \frac{jn}{k}.$$

If $j = k$, i.e. $L(\partial D) = n$, we say that ∂D is of full length.

Proof. Let $\sigma \in C_n$ be the cycle $\sigma = (0 \ \ldots \ n - 1)^{\frac{jn}{k}}$ that operates on $\mathbb{Z}_n$. We then have

$$\sigma\left(0, d_1, \ldots, \sum_{i=1}^{k} d_i\right) = \left(\frac{jn}{k}, d_1 + \frac{jn}{k}, \ldots, \sum_{i=1}^{k} d_i + \frac{jn}{k}\right)$$

$$= \left(\sum_{i=1}^{j} d_i, \sum_{i=1}^{j+1} d_i, \ldots, \sum_{i=1}^{k} d_i = 0, d_1, \ldots, \sum_{i=1}^{j-1} d_i\right),$$

$$= \left(0, d_1, \ldots, \sum_{i=1}^{k} d_i\right)$$

and thus $L(\partial D) \le \frac{jn}{k}$. As on the other hand j is minimal with $j|k$ and $d_i = d_{i+j}$ for all $1 \le i \le k - j$, $L(\partial D) \ge \frac{jn}{k}$, which proves the statement. $\qquad\square$

5.2 A centrally-symmetric $S^k \times S^1$ in $\partial\beta_{k+3}$

The following series of triangulations is due to Kühnel and Lassmann [83].

Theorem 5.6 (A centrally-symmetric $S^k \times S^1$ in $\partial\beta_{k+3}$)

There is a centrally-symmetric triangulation of $S^k \times S^1$ with $n = 2k + 6$ vertices and with a dihedral automorphism group of order $2n$. Its induced embedding into the $(k + 3)$-dimensional cross polytope is tight and PL-taut.

The construction of the triangulations is given in [83] (the triangulations are called $M_k^{k+1}(n)$ there and represented as the permcycle $[1^k 2]$). It is as follows:

Regard the vertices as integers modulo n and consider the $\mathbb{Z}_n$-orbit of the $(k + 2)$-simplex

$$\langle 0, 1, 2, \cdots, k, (k + 1), (k + 2)\rangle.$$

This is a manifold with boundary (just an ordinary orientable 1-handle), and its boundary is homeomorphic to $S^k \times S^1$. All these simplices are facets of the cross polytope of dimension $k + 3$ if we choose the labeling such that the diagonals are $\langle x, x + k + 3\rangle$, $x \in \mathbb{Z}_n$. These diagonals do not occur in the triangulation of the manifold, but all other edges are contained. Therefore we obtain a 1-Hamiltonian subcomplex of the $(k + 3)$-dimensional cross polytope. The central symmetry is the shift $x \mapsto x + k + 3$ in $\mathbb{Z}_n$. These triangulated manifolds M^{k+1} are hypersurfaces in

$\partial \beta_{k+3}$ and decompose this $(k+2)$-sphere into two parts with the same topology as suggested by the Hopf decomposition.

The same generating simplex for the group $\mathbb{Z}_m$ with $m = 2k + 5$ vertices leads to the minimum vertex triangulation of $S^k \times S^1$ (for odd k) or to the twisted product (for even k) which is actually unique [9], [33]. For any $k \geq 2$ it realizes the minimum number of vertices for any manifold of the same dimension which is not simply connected [27]. Other infinite series of triangulated sphere bundles over tori are given in [83]. Let us now come to the proof of Theorem 5.6 on the preceding page.

Proof (of Theorem 5.6*).* Define $n := 2d + 3$ and N^{d+1} as the representation of the difference cycle $(1 : 1 : \cdots : 1 : d + 1)$ over $\mathbb{Z}_n$. Then N^{d+1} is a $(d+1)$-dimensional manifold with boundary and more specific a stacked $(d+1)$-polytope with two disjoint facets identified. Thus, N^{d+1} is PL-homeomorphic to a 1-handle which is orientable if d is even and non-orientable if d is odd. The boundary $M^d := \partial N^{d+1}$ of N^{d+1} lies in Walkup's class $\mathcal{H}(d)$. Using the classification of sphere bundles from [134] we can deduce that M^d is PL-homeomorphic to $S^1 \times S^{d-1}$ if d is even and to the total space of $S^{d-1} \times S^1$, the twisted S^{d-1}-bundle over S^1, if d is odd.

It remains to show that M^d and N^{d+1} are tight triangulations. First, note that both triangulations are 2-neighborly. For $d \geq 4$, M^d and N^{d+1} lie in Walkup's class $\mathcal{H}(d) = \mathcal{K}(d)$ and thus by Theorem 3.5 on page 59, M^d and N^{d+1} are tight triangulations in this case. We will continue with an elementary proof of the tightness of M^d and N^{d+1} that also works for $d < 4$.

Note that N^{d+1} has the homotopy type of S^1 and that the generator of $\pi_1(N^{d+1})$ may be chosen as the union of all edges $\langle i, i+1 \rangle$, $0 \leq i \leq 2d + 2$. We will now show that for any subset $X \subset V(N^{d+1})$ the span of X in N^{d+1} is either contractible or homotopy equivalent to N^{d+1}. There are two cases that can be distinguished: (i) if X is contained in a $(d+2)$-tuple of subsequent vertices of N^{d+1} (i.e. a face of N^{d+1}), then X is clearly contractible, and (ii) if X is not contained in any $(d+2)$-tuple of subsequent vertices, then span(X) collapses onto a union of three edges $\langle v_1, v_2 \rangle$, $\langle v_2, v_3 \rangle$, $\langle v_3, v_1 \rangle$ where any two of the three vertices lie in a common $(d+2)$-tuple of subsequent vertices. It now follows that the union of these three edges is homotopy equivalent to the generator of $\pi_1(N^{d+1})$.

Altogether it follows that N^{d+1} is a tight triangulation. It can be shown (see [78, Prop. 6.5]) that for a tight triangulation M with $\sum \beta_i(\partial M) = 2 \sum \beta_i(M)$, the boundary ∂M is also tight. Thus, the tightness of M^d follows from the tightness of N^{d+1} as $\sum \beta_i(M^d; \mathbb{Z}_2) = 4 = 2 \sum \beta_i(N^{d+1}; \mathbb{Z}_2)$. Note that this relation does not hold for $d = 1$: N^2 is the tight 5-vertex Möbius band, but its boundary is not tight. $\square$

All triangulations M^d above have the minimal number of vertices among all triangulated d-manifolds which are not simply connected, [78, Prop. 5.7].

5.3   A conjectured series of triangulations of $S^{k-1} \times S^{k-1}$

This section contains the description of a conjectured generalization of Theorem 5.6 on page 102 with an analogous infinite series of triangulations $S^{k-1} \times S^{k-1}$ as $(k-1)$-Hamiltonian subcomplexes of $\partial \beta^{2k}$ for $k \geq 2$ with dihedral and vertex transitive automorphism group. Furthermore, it is not impossible that also infinite series of analogous triangulations of $S^k \times S^3$, $S^k \times S^5$, ... exist as Hamiltonian subcomplexes of a cross polytope, at least for odd k and again each with a dihedral and vertex-transitive group action. Note though, that the latter cases have not been investigated in the course of this work.

The existence of a $(k-1)$-Hamiltonian $S^{k-1} \times S^{k-1}$ with $n = 4k$ vertices and $d = 2k$ would give a positive answer to a conjecture by Lutz [90, p.85], and it would additionally realize equality in Sparla's inequality in Section 4.1 on page 77 for any k since

$$(-1)^{k-1}\frac{1}{2}\left(\chi(S^{k-1} \times S^{k-1}) - 2\right) = 1 = \frac{2k-1}{1} \cdot \frac{2k-3}{3} \cdot \frac{2k-5}{5} \cdot \ \cdots \ \cdot \frac{1}{2k-1}.$$

The following inductive construction seems to yield exactly the desired triangulations of $S^{k-1} \times S^{k-1}$ as $(k-1)$-Hamiltonian subcomplexes of $\partial \beta^{2k}$, but unfortunately we did not succeed in actually proving this result – so far the first cases could only be verified by hand and by computer up to $k = 12$, see Table A.4 on page 133 for some conjectured properties of the triangulations in the series.

The first two triangulations in the conjectured series, M^2 and M^4, are Altshuler's unique 8-vertex triangulation of the torus with fixed point free involution and one of the three types of Sparla's 12-vertex triangulations of $S^2 \times S^2$, respectively – see [126].

5.3.1 $(k-1)$-Hamiltonian $(2k-2)$-submanifolds of β^{2k}

Before describing the construction principle let us investigate on the f-vectors of $(k-1)$-Hamiltonian $(2k-2)$-subcomplexes of the $2k$-cross polytope. By virtue of the Dehn-Sommerville equations (see Section 1.3 on page 20), the f-vector of such complexes is completely determined by its first $(k+1)$ entries. As a $(k-1)$-Hamiltonian submanifold M of β^{2k} satisfies

$$f_i(M) = 2^{i+1}\binom{2k}{i+1} \text{ for } i \leq k-1,$$

using the Dehn-Sommerville equations for triangulated manifolds (see [56, Sect. 9.5]) we get for the number of facets of M:

$$
\begin{aligned}
f_{2k-2}(M) &= (-1)^{k-1}\binom{2k-2}{k-1}\chi(M) + 2\sum_{i=0}^{k-2}(-1)^{k+i}\binom{2k-i-3}{k-1}f_i(M) \\
&= (-1)^{k-1}\binom{2k-2}{k-1}\chi(M) + \sum_{i=0}^{k-2}(-1)^{k+i}\binom{2k-i-3}{k-1}\binom{2k}{i+1}2^{i+2}.
\end{aligned}
\tag{5.1}
$$

We will simplify the expression above. The following lemmata will prove helpful in for this.

Lemma 5.7 (R. Adin [1]) *Let K be a $(d-1)$-dimensional simplicial complex and let M be a $(k-1)$-Hamiltonian subcomplex of K. Then*

$$h_M(q) = \text{trunc}_k\left(\frac{h_K(q)}{(1-q)^{d-k}}\right),$$

where the k-truncation of a power-series is defined as

$$\text{trunc}_k\left(\sum_{i=0}^{\infty}a_iq^i\right) := \sum_{i=0}^{k}a_iq^i.$$

As this lemma that can be found in Ron Adin's PhD thesis [1, Lemma 1.5] does not seem to be published in one of his papers, the proof is reproduced along the following lines.

Proof. We will write $f_i = f_i(K)$ in the following. Note that $f_i(K) = f_i(M)$ for $i \leq k - 1$. By definition of the h-polynomial we have

$$h_K(q) = \sum_{i=0}^{d} f_{i-1} q^i (1-q)^{d-i} \quad \text{and} \quad h_M(q) = \sum_{i=0}^{k} f_{i-1} q^i (1-q)^{k-i}.$$

It follows that

$$\frac{h_K(q)}{(1-q)^{d-k}} - h_M(q) = \sum_{i=k+1}^{d} f_{i-1} q^i (1-q)^{k-i}$$

$$= q^{k+1} \sum_{j=0}^{d-k-1} f_{j+k} q^j (1-q)^{-j-1}$$

is a power series in q for which the coefficients of $1, q, \ldots, q^k$ all vanish. $\qquad\square$

In particular, this means that the first $k + 1$ entries of the h-vector of a k-Hamiltonian subcomplex M of some simplicial polytope P can be computed from the h-vector of P. More precisely we have for $0 \leq i \leq k$:

$$h_i(M) = \sum_{j=0}^{i} h_j(P).$$

This allows us to prove the following lemma.

Lemma 5.8 *Let M be a $(k-1)$-Hamiltonian $(2k-2)$-submanifold of β^{2k}. Then*

$$\chi(M) = \chi(S^{k-1} \times S^{k-1}) = 2 + 2(-1)^{k-1} = \begin{cases} 0 & \text{for even } k \\ 4 & \text{for odd } k \end{cases}. \tag{5.2}$$

Proof. As $h_i(\beta^{2k}) = \binom{2k}{i}$ and M is $(k-1)$-Hamiltonian in β^{2k}, we have by Lemma 5.7 on the previous page:

$$h_i(M) = \sum_{j=0}^{i} \binom{2k}{j}, \quad \text{for } 0 \leq i \leq k. \tag{5.3}$$

Furthermore, we have the Dehn-Sommerville equations for combinatorial $(d-1)$-manifolds with $d = 2k - 1$:

$$h_j - h_{d-j} = (-1)^{d-j}\binom{d}{j}(\chi(M) - 2), \quad \text{for } 0 \le j \le k - 1. \tag{5.4}$$

By virtue of the Dehn-Sommerville equations (5.4) we obtain

$$h_k(M) - h_{k-1}(M) = (-1)^{k-1}\binom{2k-1}{k}(\chi(M) - 2),$$

on the one hand, whereas using (5.3) on the preceding page we obtain

$$h_k(M) - h_{k-1}(M) = \binom{2k}{k}$$

on the other hand. Together this gives:

$$\chi(M) = 2 + (-1)^{k-1}\binom{2k}{k}\binom{2k-1}{k}^{-1} = 2 + 2(-1)^{k-1} = \begin{cases} 0 & \text{for even } k \\ 4 & \text{for odd } k \end{cases}. \qquad \square$$

In a similar way, we can calculate the number of facets of such a $(k-1)$-Hamiltonian $(2k-2)$-submanifold of β^{2k}. This is done in the following.

Lemma 5.9 *Let M be a $(k-1)$-Hamiltonian $(2k-2)$-submanifold of β^{2k}. Then for the number of facets of M we have*

$$f_{2k-2}(M) = 4k\binom{2k-2}{k-1}. \tag{5.5}$$

Proof. In what follows we will make use of the following three binomial identities[1]:

$$\sum_{i=0}^{k-1}(-1)^i\binom{2k-1}{i} = (-1)^{k-1}\binom{2k-2}{k-1}, \tag{5.6}$$

---

[1]The author is indebted to Isabella Novik for her kind support regarding the revision of this section of the work at hand. He wishes to thank her for the fruitful discussions on h-vectors, for pointing him to the work of Ron Adin (cf. Lemma 5.7 on page 105) and for giving hints to these binomial identities. In a first version (using f-vectors), the proofs of 5.8 on the preceding page and 5.9 were less elegant.

$$2\sum_{i=0}^{k-1}\binom{2k}{i} = 2^{2k} - \binom{2k}{k},$$ (5.7)

$$\sum_{i=0}^{k} i\binom{2k}{i} = k2^{2k-1}.$$ (5.8)

The first Identity (5.6) on the previous page can be obtained by an iteration of Pascal's identity, see 4.18 in Gould's list [54]. The second Identity (5.7) is a consequence from the fact that the sum of all binomial coefficients $\binom{2k}{i}$ is 2^{2k} and the fact that the binomial coefficients are symmetric (cf. 2.42 in Gould's list [54]). Looking at the third Identity (5.8), the left-hand-side counts the number of subsets of $\{1,\ldots,2k\}$ with size $i \leq k$ with a distinguished element. For each such subset, we can first choose its distinguished element for which there are $2k$ choices and then complete it to a set of size i, which amounts to choosing $i-1 \leq k-1$ elements out of a $(2k-1)$-element set for which there are exactly 2^{2k-2} choices, which shows (5.8).

Let us now continue with the proof of Lemma 5.9 on the previous page. The number of facets of M in terms of the h-vector is

$$f_{2k-2}(M) = \sum_{i=0}^{2k-1} h_i(M).$$

Using the Dehn-Sommerville equations (5.4) on the preceding page, the above sum can be written in terms of h_i, $i \leq k-1$ and $\chi(M)$:

$$f_{2k-2}(M) = 2\sum_{i=0}^{k-1} h_i(M) + \sum_{j=0}^{k-1}(-1)^j\binom{2k-1}{j}(\chi(M)-2).$$

Using the identities (5.2), (5.3) and (5.6), the above transforms to

$$f_{2k-2}(M) = 2\sum_{i=0}^{k-1}\sum_{j=0}^{i}\binom{2k}{j} + \underbrace{\sum_{j=0}^{k-1}(-1)^j\binom{2k-1}{j}}_{=(-1)^{k-1}\binom{2k-2}{k-1}} \cdot (\chi(M)-2) = (*)$$
$$\underbrace{\qquad\qquad\qquad}_{=(-1)^{k-1}\binom{2k-2}{k-1}\,\cdot\,2(-1)^{k-1}}$$

$$(*) = \sum_{i=0}^{k-1} 2(k-i)\binom{2k}{i} + 2\binom{2k-2}{k-1},$$

$$= 2k\sum_{i=0}^{k-1}\binom{2k}{i} - 2\sum_{i=0}^{k-1} i\binom{2k}{i} + 2\binom{2k-2}{k-1}.$$

From this we obtain

$$f_{2k-2}(M) = k\left(2^{2k} - \binom{2k}{k}\right) - \left(k2^{2k} - 2k\binom{2k}{k}\right) + 2\binom{2k-2}{k-1},$$

$$= k\binom{2k}{k} + 2\binom{2k-2}{k-1},$$

$$= (4k-2)\binom{2k-2}{k-1} + 2\binom{2k-2}{k-1},$$

$$= 4k\binom{2k-2}{k-1},$$

as claimed, where the identities (5.7) and (5.8) were used in the first step. □

Moreover, there is the following lower bound theorem for centrally symmetric triangulations of $2k$-manifolds due to Sparla [126].

Theorem 5.10 (Lower Bound Theorem, Th. 3.6 in [124])
Let M be a combinatorial $2k$-manifold that contains all vertices of a centrally symmetric simplicial polytope $P \subset E^d$. If $\mathrm{skel}_k(P) \subset \mathrm{skel}_k(|M|)$ and M is a subcomplex of $C(\partial P)$, then the following holds:

$$(-1)^k\binom{2k+1}{k+1}(\chi(M)-2) \geq 4^{k+1}\binom{\frac{1}{2}(d-1)}{k+1}. \tag{5.9}$$

For $d > 2k+1$ equality in (5.9) holds if and only if P is affinely equivalent to the cross polytope β^d.

We will show in the course of this chapter that the above inequality is sharp for $k \leq 12$. We furthermore conjecture that it is sharp for all k.

5.3.2 The construction principle Φ

Let M^2 be Altshuler's unique centrally symmetric triangulation of the 2-torus $T = S^1 \times S^1$ on 8 vertices [126, Satz A.1] given by the following facet list:

$$
\begin{array}{llll}
\langle 012 \rangle, & \langle 017 \rangle, & \langle 025 \rangle, & \langle 035 \rangle, \\
\langle 036 \rangle, & \langle 067 \rangle, & \langle 123 \rangle, & \langle 136 \rangle, \\
\langle 146 \rangle, & \langle 147 \rangle, & \langle 234 \rangle, & \langle 247 \rangle, \\
\langle 257 \rangle, & \langle 345 \rangle, & \langle 456 \rangle, & \langle 567 \rangle.
\end{array}
\tag{5.10}
$$

M^2 is a 1-Hamiltonian subcomplex of β^4 assuming a vertex labeling of the cross polytope with elements of $\mathbb{Z}_8$ such that the four diagonals are the edges $\langle i, i+4 \rangle$, $0 \leq i \leq 3$. The full automorphism group G^2 of M^2 is generated by the three permutations

$$
\sigma = (0,1,2,3,4,5,6,7), \quad \tau = (0,6)(1,5)(2,4), \quad \lambda = (0,2)(1,5)(4,6).
$$

G^2 is of order 32 and can be written as a semi direct product $(D_8 \times C_2) \rtimes C_2$, where the leftmost cycle σ above generates the cyclic subgroup C_n and the two multiplications $\tau : x \mapsto -x \mod n$ and $\lambda : x \mapsto 3x \mod n$ represent the two elements of order 2 above and correspond to two subgroups of type C_2.

If one now takes the link of the vertex 0 in M^2 and expands this link by the procedure Φ as explained below, one gets a new 4-dimensional combinatorial manifold M^4 that is a Hamiltonian subcomplex of β^6, which topologically is a sphere product $S^2 \times S^2$ and has an automorphism group of the same type, see Remark 5.12 on page 113.

Construction Principle Given the complex M^d, one can construct a new simplicial complex M^{d+2} from M^d by the following procedure Φ:

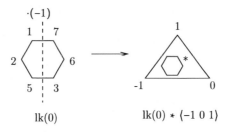

Figure 5.1: Construction of $M^4 \cong S^2 \times S^2$ from the link of 0 in $M^2 \cong S^1 \times S^1$.

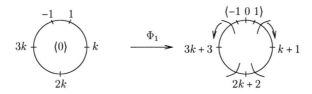

Figure 5.2: Construction of $M^{2k+2} \cong S^{k+1} \times S^{k+1}$ from the link of 0 in $M^{2k} \cong S^k \times S^k$.

First, the vertex labels of the link $L_0 := \mathrm{lk}_{M^d}(0)$ of the vertex 0 in M^d are embedded into $\mathbb{Z}_{4k+4}$ as follows:

$$\psi: \ \mathbb{Z}_{4k} \setminus \{\overline{0}, \overline{2k}\} \ \rightarrow \ \mathbb{Z}_{4k+4}$$
$$\overline{v} \ \mapsto \ \begin{cases} \overline{v+1} & \text{for } 1 \le v \le 2k-1 \\ \overline{v+3} & \text{for } 2k+1 \le v \le 4k-1 \end{cases}.$$

In succession, the join of all simplices in the link L_0 with a new simplex $\langle -1, 0, 1 \rangle = \langle 4k+3, 4k+4, 4k+5 \rangle$ is taken, again in $\mathbb{Z}_{4k+4}$. This yields a new simplicial complex $\tilde{M}^{d+2}$.

Note that if M^d did not contain any diagonal $\langle i, 2k+i \rangle$ as face, then $\tilde{M}^{d+2}$ does not contain any element of the form $\langle i, 2(k+1)+i \rangle$. These missing edges will be the diagonals of the complex $\tilde{M}^{d+2}$.

In a last step, the group operation of a group G generated by the permutations

$$\sigma = (0, \ldots, 4k+3), \quad \tau = (1, 4k+3)(2, 4k+2)\ldots(2k+1, 2k+3), \tag{5.11}$$

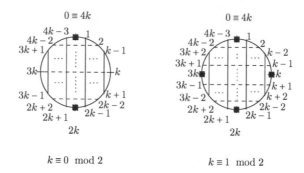

Figure 5.3: Multiplication $\lambda : v \mapsto (2k-1)v \mod 4k$ with fixed points shown as black squares. Geometrically, λ reflects odd vertices along the horizontal axis and even vertices along the vertical axis. For even k (left side) λ has two fixed points, for odd k (right side) it has four.

and, depending on the parity of k, one of the two permutations

$$\lambda = \begin{cases} (3k-1, 3k+1)(3k-3, 3k+3)\dots(2k+1, 4k-3) \\ \cdot(2k-1, 1)(2k-3, 3)\dots(k+1, k-1) & \text{for even } k \\ \cdot(2, 4k-2)(4, 4k-4)\dots(2k-2, 2k+2) \\ \\ (3k-2, 3k+2)(3k-4, 3k+4)\dots(2k+1, 4k-3) \\ \cdot(2k-1, 1)(2k-3, 3)\dots(k+2, k-2) & \text{for odd } k \\ \cdot(2, 4k-2)(4, 4k-4)\dots(2k-2, 2k+2) \end{cases} \tag{5.12}$$

on the set of simplices in $\tilde{M}^{d+2}$ is considered. σ is a rotation and τ, λ correspond to multiplications $\tau : v \mapsto -v \mod 4k$ and $\lambda : v \mapsto (2k-1)v \mod 4k$, respectively. See Figure 5.3 for an illustration of the operation of the multiplication λ on the set of vertices. Note that this group operation leaves the diagonals of $\tilde{M}^{d+2}$ invariant and we have the following result.

Corollary 5.11

If M^{2k}, $k \geq 2$ is a complex obtained from M^2 by iterating the process Φ, then M^{2k} contains $2k$ diagonals $\langle i, 2k+i \rangle$, $0 \leq 1 < 2k$, and thus is a subcomplex of the $2k$-cross polytope β^{2k}.

5.12 Remark *Sparla showed in [124, Th. B.1] that there exist exactly three combinatorial types of centrally symmetric triangulations of $S^2 \times S^2$ on twelve vertices, all of which are 2-Hamiltonian subcomplexes of $C(\partial \beta^6)$. The complex M^4 obtained by the procedure Φ from the link of the vertex 0 in M^2 corresponds to the type M_3 in [124, Th. B.1].*

Now we can state the central conjecture of this chapter.

Conjecture 5.13

The manifolds M^{2k-2} obtained by the construction principle Φ above are k-Hamiltonian submanifolds of β^{2k} for any k. Topologically, they are triangulations of $S^{k-1} \times S^{k-1}$.

We can at least prove Conjecture 5.13 up to a value of $k = 12$: Up to $k = 12$, the construction Φ was carried out on a computer using the software package `simpcomp` [44, 45] and the correctness of Conjecture 5.13 was verified as described below.

Theorem 5.14

For $k \leq 12$ there exist centrally symmetric triangulations of $S^{k-1} \times S^{k-1}$ that are k-Hamiltonian subcomplexes of β^{2k}.

The proof uses the following theorem due to Matthias Kreck published in [72].

Theorem 5.15 (M. Kreck [72])

Let M be a 1-connected smooth codimension 1 submanifold of S^{d+1} and $d > 4$. If M has the homology of $S^k \times S^{d-k}$, $1 < k \leq \frac{d}{2}$, then M is homeomorphic to $S^k \times S^{d-k}$. If $d = 4$, then M is diffeomorphic to $S^2 \times S^2$. The corresponding statement holds for PL respectively topological manifolds, if one replaces "smooth" by "PL", respectively "topological" and "diffeomorphic" by "PL-homeomorphic", respectively "homeomorphic" in the preceding statement.

Proof. First of all, S^{d+1} decomposes by $X \cup Y$ by the generalized Jordan separation theorem [123], X, Y are 1-connected by the Seifert-van-Kampen theorem [138, Thm. 2.5] and have the homology of S^k and S^{d-k}, respectively. As $k \leq \frac{d}{2}$, the generator of $\pi_k(X) \cong H_k(X)$ can be represented by an embedding of S^k in the interior of X by the Hurewicz theorem [138, Thm. 7.1]. Now denote by E the normal bundle of S^k in S^{d+1} and choose a tubular neighborhood to identify the disk bundle of E with a neighborhood of S^k in X. In a next step, it has to be verified that the complement C of the interior of the disk bundle in X is an h-cobordism between the sphere bundle and $\partial X = M$. For $d > 4$, the h-cobordism theorem [100] implies that M is diffeomorphic to the sphere bundle of E. If $d = 4$, one can use Freedman's topological h-cobordism theorem [49] to conclude that M is homeomorphic to the sphere bundle of E.

It remains to verify that C is an h-cobordism. By virtue of the Seifert-van-Kampen theorem C is 1-connected and by virtue of the Mayer-Vietoris sequence the inclusions of the sphere bundle of E and from ∂X to C induce isomorphisms in the homology up to dimension $\frac{d}{2}$. By Lefschetz duality the inclusion then also induces isomorphisms on the remaining homology groups and by the Whitehead theorem [138, Thm. 7.13] both inclusions are homotopy equivalences.

To finish the proof, we show that the bundle E is the trivial bundle. As E is the stable normal bundle of S^k in S^{d+1}, it is stably trivial. Now as $k \leq \frac{d}{2}$, the dimension of the vector bundle E is larger than k and in this case a stably trivial bundle has to be trivial.

Note that since all tools used above are also available for PL and topological manifolds by the fundamental work of Kirby and Siebenmann [69], the corresponding statement also holds for PL respectively topological manifolds M, where the term "diffeomorphism" above has then to be replaced by the term "PL-isomorphism" or "homeomorphism", respectively. □

Let us now come to the proof of Theorem 5.14 on the preceding page.

Proof (of 5.14). For $k \leq 3$, the statement was known to be true before [124]. For $k \geq 4$, Kreck's Theorem 5.15 on the previous page can be applied and states that we only have to verify that the complexes are combinatorial manifolds and have the homology of $S^{k-1} \times S^{k-1}$, i.e. that

$$H_i(M^{2k-2}) = \begin{cases} \mathbb{Z} & \text{for } i = 0, 2k, \\ \mathbb{Z}^2 & \text{for } i = k, \\ 0 & \text{otherwise.} \end{cases}$$

For the complexes M^{2k-2} obtained by the process Φ this is indeed the case for $k \leq 12$, as was checked with the help of a computer algorithm modeling the process Φ in the GAP system with the help of the package simpcomp, see Appendix C on page 139 and Appendix E on page 177. □

As a consequence of Theorem 5.14 on page 113 we have the following result.

Corollary 5.16

Sparla's inequality (5.9) on page 109 for combinatorial $(k-1)$-connected $2k$-submanifolds M that are k-Hamiltonian subcomplexes of the $(2k+2)$-cross polytope is sharp up to $k = 12$.

Note that the only values for $\chi(M)$ that can occur above are $\chi = 0$ for odd k and $\chi = 4$ for even k, see Corollary (5.2) on page 106. We furthermore conjecture (5.9) to be sharp for all values of k.

Conjecture 5.17

For each k, the triangulation M^{2k} as constructed in Section 5.3.2 fulfills equality in (5.9) on page 109. In particular, Sparla's inequality (5.9) for combinatorial $(k-1)$-connected $2k$-submanifolds that are k-Hamiltonian subcomplexes of the $(2k+2)$-cross polytope is sharp for all k.

5.3.3 The construction Φ using difference cycles

In order to facilitate the construction Φ, we will mod out the operation of the cyclic group (as a subgroup of the full automorphism group) in the representation of the triangulations M^d. This allows us to work on the level of difference cycles. Note that —as before— we will work with difference cycles that are not necessarily given in their minimal representation in the following. In terms of difference cycles, the anchor point of the iterative process, the triangulation M^2, can be written as

$$(1:1:6), \ (3:3:2), \tag{5.13}$$

where the two difference cycles are of full length and encode the facet list shown in (5.10) on page 110.

In order to show how the process Φ can also be explained on the level of difference cycles, we have to do two things. First, we have to establish the join process with the simplex $\langle -1, 0, 1 \rangle$ and in a second step we have to explain a multiplication on the set of difference cycles, see Definition 5.2 on page 100. This will be discussed in the following. As a result, counting the number of facets of the complexes M^d is facilitated.

The join with the simplex $\langle -1, 0, 1 \rangle$ can be carried out on a difference cycle level and corresponds to the operation of gluing a sequence of the form $1 : 1$ into the difference cycle, while simultaneously increasing the "opposite" difference by 2 to accommodate for the change of modulus from $4k$ to $4k + 4$. Let us explain what we mean by "opposite" here. Since none of the simplices may contain one of the diagonals $\langle i, i + 2k \rangle$ as a face, for any representative $\delta B = (d_1, \ldots, d_k)$ of a difference cycle ∂D and for any entry d_i of δB, there exists no entry d_j of δB such that there exits a *running sum* (see Definition 5.18 on the facing page) $\sum_i^j = 2k$. Thus, there exits an entry d_j with a minimal index j such that $\sum_i^j > 2k$. This entry d_j is referred to as the "opposite entry of d_i". Note that necessarily $d_j > 1$ must hold here.

We refer to the join procedure with the simplex $\langle -1, 0, 1 \rangle$ as *inheritance* in the following sense: if a difference cycle ∂C of dimension $2k$ yields a new difference cycle ∂D of dimension $2k$ under the process Φ, then ∂C will be referred to as *father cycle* and ∂D as *child cycle*. The inheritance process here takes place on diagonals of the triangle structure shown in Table A.1 on page 130 and Table A.2 on page 131 and we will also write $\partial C \searrow \partial D$ to state the fact that ∂D is a child cycle of the father cycle ∂C. The superscript in the inheritance symbol $\searrow$ and more details of the inheritance scheme will be explained in Section 5.3.4 on page 120.

Let us now describe a multiplication on difference cycles. Note that for ease of notation we will sometimes just write D instead of ∂D for a difference cycle from now on, if no confusion can be expected.

Definition 5.18 (running sum) *Let $k, n \in \mathbb{N}$, $n > k$, and let $D = (d_0 : \cdots : d_k)$ be a difference cycle in $\mathbb{Z}_n$. Then a* running sum *of D with value S is a sum of* *consecutive entries of D of the form*

$$\sum_{i=l}^{l+m} d_i = S,$$

where $m \leq k$, the summation is carried out over $\mathbb{Z}$ and the index i is taken modulo n. For ease of notation we will most of the time just write $\Sigma_l^{l+m}(D)$ or just Σ_l^{l+m} from now on.

This allows us to reformulate the facts of the process Φ established above on the level of difference cycles.

Corollary 5.19

If a difference cycle D does not contain any of the diagonals, i.e. a running sum S with $S \equiv 2k \mod 4k$, then the set of all difference cycles obtained from D by the construction Φ does not contain any difference cycle $\hat{D}$ with a running sum $S \equiv 2(k+1) \mod 4(k+1)$.

This means that the complexes obtained by the process Φ are subcomplexes of the $2k$-cross polytope for each k.

The next lemma will help us to better understand the structure of the difference cycles obtained with the process Φ.

Lemma 5.20 *Let $D = (d_1 : \cdots ; d_{2k-1})$ be a difference cycle over $\mathbb{Z}_{4k}$ with an even number of running sums S with $S \equiv 1 \mod 4k$ and let $p(D)$ and $q(D)$ denote the number of distinct running sums of D with $\Sigma_a^b(D) \equiv 1 \mod 4k$ and $\Sigma_a^b(D) \equiv 2k-1 \mod 4k$, respectively. Furthermore, let $\hat{D} := (2k-1) \cdot D$. Then the following holds:*

$$p(\hat{D}) = q(D) \text{ and } q(\hat{D}) = p(D).$$

Proof. If there exist entries $d_i \equiv 1$ or $d_i \equiv 2k-1$ in D, these get mapped to entries $\hat{d}_i = 2k-1$ or $\hat{d}_i = 1$ in $\hat{D}$, respectively. We will thus exclude these cases from now on and look at running sums $\Sigma_a^b(D) > 2k-1$. On the one hand, $(2k-1) \cdot \Sigma_a^b = ((2k-1)d_a : \cdots : (2k-1)d_b) \equiv 2k-1 \mod 4k$ and analogously for any $\Sigma_a^b(D) \equiv 2k-1$

117

mod $4k$ one has $(2k-1)\cdot\sum_a^b \equiv 1 \mod 4k$, i.e. $p(\hat{D}) \geq q(D)$ and $q(\hat{D}) \geq p(D)$. On the other hand, if there exists a running sum $(2k-1)\sum_a^b \equiv 2k-1 \mod 4k$, then $\sum_a^b \equiv 1 \mod 4k$ and if there exists a running sum $(2k-1)\sum_a^b \equiv 1 \mod 4k$, then $\sum_a^b \equiv 2k-1 \mod 4k$, as $\gcd(2k-1,4k) = 1$ and $(2k-1)^2 \equiv 1 \mod 4k$, i.e. $p(\hat{D}) \leq q(D)$ and $q(\hat{D}) \leq p(D)$. Thus, $p(\hat{D}) = q(D)$ and $q(\hat{D}) = p(D)$. $\qquad\square$

Now we can make another observation on the difference cycles obtained by the process Φ.

Lemma 5.21 *Let D be a difference cycle of length $2k-1$ that is obtained by the process Φ and let $p(D)$, $q(D)$ be defined like in Lemma 5.20 on the preceding page. Then the following holds:*

$$p(D) + q(D) = 2k - 2. \qquad (5.14)$$

Particularly, all difference cycles obtained by the process Φ have an even number of differences of the form $d_i = 1$ and an even number of running sums with value $2k-1$.

Proof. The statement follows by induction. It holds for $k = 2$ and the two difference cycles $(1:1:6)$ (with $p = 2$ and $q = 0$) and $(3:3:2)$ (with $p = 0$ and $q = 2$). Now assume that the statement holds for k. A difference cycle $D = (d_1 : \cdots : d_{2k-1})$ over $\mathbb{Z}_{4k}$ for which (5.14) holds is mapped to a difference cycle $\hat{D} = (\hat{d}_1 : \cdots : \hat{d}_{2k+1})$ over $\mathbb{Z}/(4k+4)\mathbb{Z}$ by the mapping Φ. We will now show that if $p(D) + q(D) = 2k - 2$, then $p(\hat{D}) + q(\hat{D}) = 2(k+1) - 2 = 2k$. Obviously, $p(\hat{D}) = p(D) + 2$ holds. It thus remains to show that $q(\hat{D}) = q(D)$. We will first show that $q(D) \geq q(\hat{D})$, i.e. that any running sum $\hat{S} = \sum_{i=a}^b \hat{d}_i = 2k+1$ in $\hat{D}$ implies that there exists a running sum $S = \sum_{i=\tilde{a}}^{\tilde{b}} d_i = 2k-1$ in D. Since $\sum \hat{d}_i - \sum d_i = 4$ and the differences of $\hat{D}$ differ from the ones of D only by a consecutive pair of differences with $\hat{d}_i = \hat{d}_{i+1} = 1$ and one "opposite" entry $\hat{d}_j = d_k + 2$ for some k, either the pair of differences $\hat{d}_i = \hat{d}_{i+1} = 1$ or the "opposite" entry are part of the running sum. This means that if there exists a running sum $\hat{S}$ in $\hat{D}$ like above, then there also exists a uniquely corresponding running sum S like above in D, i.e. $q(D) \geq q(\hat{D})$. On the other hand every running sum $S = 2k-1$ in D yields a running sum $\hat{S} = 2k+1$ in $\hat{D}$ as it contains exactly on pair of antipodal points of $\mathbb{Z}_{4k}$, i.e. $q(D) \leq q(\hat{D})$. Altogether it follows that $q(D) = q(\hat{D})$ which finishes the proof of the statement. $\qquad\square$

Thus, the set of difference cycles obtained by the process Φ for each k can be grouped by the number of running sums S_i with $S_i \equiv 1 \mod 4k$ and $S_i \equiv (2k-1)$ mod k. We will refer to these grouped difference cycles as *classes of difference cycles* and denote a class of difference cycles on level k (i.e. the difference cycles are of dimension $2k-2$) that contains j distinct entries $d_i = 1$ by the symbol $\mathcal{C}^k_j$, see Table A.1 on page 130 for the classes and Table A.2 on page 131 for the conjectured number of class elements. The latter form a Pascal triangle scheme, and in fact one seems to obtain a squared Pascal triangle, i.e. Pascal's triangle where each entry is squared. For example, the top element of the triangle in Table A.2 on page 131 for $k = 1$ is the class C^1_0 that contains one difference cycle of the form $\partial D = (4)$. Note that the inheritance relation $\searrow$ maps a father cycle in $\mathcal{C}^k_j$ to child cycles in $\mathcal{C}^{k+1}_{j+2}$ as described before.

The multiplication $\cdot \lambda_k = 2k - 1$ operates on the set of classes C^k_j for each fixed k.

Corollary 5.22

The multiplication

$$\lambda_k : \quad \mathbb{Z}_{4k} \to \mathbb{Z}_{4k}$$
$$\overline{x} \mapsto (2k-1)\overline{x} \mod 4k$$

acts as an involution on the set of difference cycle classes, mirroring the difference cycles along the vertical axis of the squared Pascal triangle shown in Table A.1 on page 130 and Table A.2 on page 131. For any difference cycle D, $q(D) = p(\lambda \cdot D)$ and $p(D) = q(\lambda \cdot D)$ holds.

Thus, the operation of the multiplication λ_k yields orbits of length two in the general case — but there exist special cases for odd k where difference cycles have $(2k-1)$ as multiplier, see Definition 5.1 on page 99. These form 1-element orbits of the operation $\cdot \lambda_k$ and for those difference cycles necessarily $p(D) = q(D)$ holds. This will be of interest in the next section where we will have a closer look at the class cardinalities shown in Table A.2 on page 131.

The second multiplication $\cdot(-1)$ acts as an involution on the set of difference cycles by reversing their entries. See Table 5.1 on page 121 for the number of difference cycles in M^{2k-2}, $1 \leq k \leq 7$, that have the multiplier $\cdot(-1)$, again grouped by the number of running sums S_i with $S_i \equiv 1 \mod 4k$ and $S_i \equiv (2k-1) \mod k$. These seem to form a Pascal triangle, too.

Comparing the conjectured number of difference cycles in Table A.2 on page 131 and taking the row sum for every k, we obtain

$$\sum_{i=0}^{k-1} \binom{k-1}{i}^2 = \binom{2k-2}{k-1}$$

as the conjectured number of difference cycles in each complex M^{2k-2}. Assuming furthermore orbits of full length under the operation of the cyclic automorphism group (see Lemma 5.5 on page 101), we obtain

$$f_{2k-2}(M^{2k-2}) = 4k\binom{2k-2}{k-1}$$

as the conjectured number of facets of the complex M^{k-2}. Lemma 5.9 on page 107 tells us that these are the facet numbers that a k-Hamiltonian $(2k-2)$-manifold in β^{2k} must have. Since the complexes M^{2k-2} are k-Hamiltonian in β^{2k}, we know that $\tilde{\beta}_0 = \cdots = \tilde{\beta}_{k-1} = 0$. Assuming that M^{2k-2} is an orientable manifold, we get $\tilde{\beta}_{2k-3} = \cdots = \tilde{\beta}_{k+1} = 0$ by Poincaré duality. As by Lemma 5.8 on page 106 the Euler characteristic of M^{2k-2} is $\chi = 2 + 2(-1)^{k-1}$, this determines the "genus" of M^{2k-2} as $\frac{\beta_k}{2} = 1$. For $k \geq 4$, we can now apply Kreck's Theorem 5.15 on page 113: M^{2k-2} is a manifold with Euler characteristic $\chi(M^{2k-2}) = 2 + 2(-1)^{k-1}$ and thus has the same homology as $S^{k-1} \times S^{k-1}$. Altogether it now follows that

$$M^{2k-2} \cong_{\mathrm{PL}} S^{k-1} \times S^{k-1}.$$

5.3.4 More on inheritance

Let us now sketch how one could proof that the process Φ indeed yields the number of orbits as shown in Table A.2 on page 131. The key ingredient here is to understand how the orbits behave under multiplication with $\lambda_k = 2k - 1$ and how the process of inheritance works. Remember that in this terminology the $(2k-1)$-difference cycles yielding new $(2k+1)$-difference cycles with an appended sub-sequence $1 : 1$,

Table 5.1: The conjectured number of difference cycles of the triangulations of $M^{2k-2} \cong S^{k-1} \times S^{k-1}$ as Hamiltonian subcomplexes of β^{2k} for $k = 2, \ldots, 7$ that have the multiplier $\cdot(-1)$.

$k = 1$:					1			
$k = 2$:				1		1		
$k = 3$:			1		2		1	
$k = 4$:		1		3		3		1
$k = 5$:	1		4		6		4	1
$k = 6$: 1	5		10		10		5	1
$k = 7$: 1	6		15		20		15	6	1

the former are called father cycles and the latter are referred to as child cycles. The inheritance process is along the diagonals of the triangle structure shown in Table A.2.

First of all note that for every difference cycle ∂D, every of its entries $d_i \neq 1$ is an opposite entry for some other element (or a sequence of elements of type $1 : \cdots : 1$) of ∂D. This means that under the process Φ every father cycle has as many distinct child cycles as it has entries $d_i \neq 1$ and that every child cycle has at least one sub-sequence of elements of the form $(d_i : d_{i+1}) = 1 : 1$. Cycles that do not posses this property are called *fatherless* or *orphan* as they do not have a father cycle.

Looking at the orbit schema as shown in Table A.2 and keeping in mind that the multiplication with $\lambda_k = 2k - 1$ "mirrors" the cycles along the middle of the triangle structure in the sense of Lemma 5.21 on page 118, it is obvious that for every k the one cycle counted by the rightmost 1-entry of the triangle row is obtained by a multiplication of the cycle counted by the leftmost 1-entry of the row by $\lambda_k = 2k - 1$. The leftmost diagonal of the triangle in Table A.2 contains only 1-entries that are children of each other. We will show this by induction. For the (degenerate) case of $k = 1$ the only difference cycle is the difference cycle consisting of one difference, $\partial D_1 = (4)$. This cycle is invariant under the multiplication $\cdot \lambda_k = 2k - 1 = 1$ and $\cdot(-1)$ in $\mathbb{Z}_{4k}$ (compare Table 5.1). In terms of inheritance, the cycle $\partial D_1 = (4)$ gives birth to the cycle $\partial D_2 = (1 : 1 : 6)$ as described before. ∂D_2 in term gives birth to

121

the cycle $\partial D_3 = (1:1:1:1:8)$ and so on, with $\partial D_k = (1:\cdots:1:2k+2)$. Note that these cycles only have one entry $d_i \neq 1$ and that they just have one child under the process Φ, as a difference of 1 can never be opposite to any other difference as the cycles must not allow $2k$ as a running sum. Secondly, the cycles obtained in this way are all invariant under the multiplication of $\cdot(-1)$, i.e. the difference cycles have -1 as a multiplier, compare Table 5.1. Now as the difference cycles counted by the rightmost 1-elements of Table A.2 are obtained from the cycles ∂D_i by multiplication with $\lambda_k = 2k-1$, these have no 1-entries and thus yield $2k-1$ distinct children under the process Φ. Since the property of the maximal number of children a cycle can have under the process Φ is constant on the diagonals of the triangle in Table A.2, each diagonal can be assigned its "fertility number" in terms of this maximal number of children a cycle can have under the process Φ.

Using this inheritance scheme one can try to show that the process Φ indeed yields the cycle numbers claimed. This has been done for the first diagonal already (see above) and will be shown for the second diagonal in the following. Unfortunately, the proof in its full generality has to be left open here as there exist cycles in the process that are not children of any other cycles, the numbers of which have to be known in order to complete the proof. These orphans have $\lambda_k = 2k-1$ as multiplier and can only appear for odd values of k as they have the same number of 1-entries and distinct running sums with value $2k-1$. These will be closer investigated upon in the following. The general proof could then be carried out in a double induction on the diagonals and the elements of the diagonals of the triangle shown in Table A.2.

5.3.5 Counting difference cycles and inheritance

Let us describe the inheritance scheme $\searrow$ of the process Φ as illustrated in Table A.1 and Table A.2 in more detail. One particularity of this inheritance is that one difference cycle ∂D can be obtained by the process Φ from different father difference cycles that are not equivalent, i.e. there occur situations where two distinct difference cycles $\partial F_1 \neq \partial F_2$ have a common child, i.e. that there exists a ∂D with $\partial F_1 \searrow \partial D$

and $\partial F_2 \searrow \partial D$. This has to be taken into account when counting the number of difference cycles obtained by the inheritance $\searrow$.

Definition 5.23 (counting value of a difference cycle) *Let ∂D be a difference cycle that is obtained by the process Φ. Then we define a counting value or valuation of ∂D by the map*

$$v(\partial D) := 1 + \frac{f_{max} - f}{f+1} = \frac{f_{max} + 1}{f+1} \in \mathbb{Q},$$

where f denotes the number of fathers of ∂D (i.e. the number of distinct subsequences $d_i : \cdots : d_{i+j}$ of ∂D with $d_i = \cdots = d_{i+j} = 1$), f_{max} the maximal number of fathers of ∂D which equals the maximal number of children that can occur in the diagonal that ∂D belongs to.

This valuation is motivated by the following law that the inheritance adheres to.

Inheritance Rule: Let ∂D be a difference cycle with $f > 0$ fathers lying in the class $\mathcal{C}_1$ and assume that $\mathcal{C}_1 \searrow \mathcal{C}_2$. Then ∂D has v children with v fathers and $f_{max} - f$ children with $v + 1$ fathers. As we want to avoid double counting children, we will for each child that has several fathers only attribute a fraction of the child to each father cycle. Eliminating double counting, ∂D thus has one child with f fathers and $\frac{f_{max} - f}{f+1}$ children with $f + 1$ fathers. This is reflected in the valuation, i.e. $v(c)$ counts the (fractional) amount of children that ∂D contributes to the class $\mathcal{C}_2$ and $v(\mathcal{C}_1) = \#\mathcal{C}_2$. In the special case that ∂D is fatherless, it will have f_{max} children with one father, i.e. $v(\partial D) = C$ is this case.

It will be shown in the following that using the valuation function v, the number of children yielded by one class of difference cycles of the process Φ can be calculated and thus that the inheritance $\searrow$ yields the number of difference cycles claimed in Table A.2.

Remember that the Pascal triangle like structure of Table A.2 is symmetric to the central vertical axis by virtue of the multiplication with $\lambda_k = 2k - 1$ as was described already. Thus, if the number of children produced by one class of difference cycles is known, this yields one new entry in the triangle of Table A.1. The number of difference cycles in the class that is obtained from the newly obtained class by the multiplication with λ_k is also known, as all orbits of the operation $\cdot\lambda_k$ must have length two, unless k is odd. In the latter case there exists a "middle class" that is invariant under the operation of $\cdot\lambda_k$ and in this case fatherless or orphan difference cycles can occur constituting one-element orbits of $\cdot\lambda_k$. These difference cycles contain no sub-sequences of the type $(1 : \cdots : 1)$ of length 2 or longer and have λ_k as a multiplier.

The orphan difference cycles play an important role for the inheritance scheme of Φ as will be shown in the following.

We will now show that the operation Φ does yield the number of difference cycles as shown in Table A.2, at least for the first two diagonals $\searrow^1$ and $\searrow^3$. The general case could be proved (if one knew the exact number of orphan difference cycles for all k) using a double induction: in the inner induction the claimed number of children in each step is proved for one fixed diagonal $\searrow^i$, where the outer induction runs over all diagonals, where the information obtained in the steps before has to be used (via the mirroring operation given by the multiplication $\cdot\lambda_k$).

5.3.6 Putting it all together

As was shown before, the classes $\mathcal{C}^k_{2k-2}$ in the first diagonal of Table A.1 on page 130 are all of cardinality 1. By multiplication with λ_k, the same holds for the classes $\mathcal{C}^k_0$. Let us now have a look at the inheritance $\searrow^3$ on the second diagonal. Here we start with the single member $(3:3:2)$ of $\mathcal{C}^2_0$ that has no 1-entries and thus has 3 child cycles $\partial A = (3+2:3:1:1:2) = (1:1:2:5:3)$, $\partial B = (3:3:1:1:3:2+2) = (1:1:3:4:3)$ and $\partial C = (1:1:3:3+2:2) = (1:1:3:5:2)$ with one father each. The class $\mathcal{C}^3_2$ is invariant under multiplication with $\lambda_k = 5$. The cycles ∂A and ∂B have λ_k as multiplier and the cycle ∂B gets mapped to $\partial D := \lambda_k \partial B = (1:2:1:4:4)$, a fatherless cycle. Counting the child cycles obtained by applying the valuation function of Section 5.3.5 on page 122 and denoting the sets of difference cycles with i father cycles by F_i, we successively get the element count of the classes $\mathcal{C}^k_{2k-4}$, using the valuation function $v = \frac{f_{max}-f}{f+1}$ with $f_{max} = 7$ and denoting above the arrows the fraction of distinct child elements that one difference cycle yields under inheritance:

k	$\#F_0$	$\#F_1$	$\#F_2$	$\#F_3$	$\Sigma \#F_i$	$\binom{k-1}{2}^2$
3	1	3	0	0	4	4
		$\cdot\frac{3}{1}$ + $\cdot\frac{2}{2}$ + $\cdot\frac{1}{3}$ +				
4	0	6	3	0	9	9
		$\cdot\frac{3}{1}$ + $\cdot\frac{2}{2}$ + $\cdot\frac{1}{3}$ +				
5	0	6	9	1	16	16
		$\cdot\frac{3}{1}$ + $\cdot\frac{2}{2}$ + $\cdot\frac{1}{3}$ +				
6	0	6	15	4	25	25
		$\cdot\frac{3}{1}$ + $\cdot\frac{2}{2}$ + $\cdot\frac{1}{3}$ +				
7	0	6	21	9	36	36
		$\cdot\frac{3}{1}$ + $\cdot\frac{2}{2}$ + $\cdot\frac{1}{3}$ +				
8	0	6	27	16	49	49

$$\qquad (5.15)$$

$$\cdots$$

Note that the special case of fatherless cycles does not appear in the steps after $k = 3$. Writing for each step $k > 4$: $a_k = \#F_1$, $b_k = \#F_2$ and $c_k = \#F_3$, we get:

$$a_k = 6, \ b_k = 3 + 6k \text{ and } c_k = \frac{1}{3}b_k - 1 + c_{k-1}.$$

It is

$$c_k = \frac{1}{3}\left(3 + 6(k-1) + \sum_{i=0}^{k-2} b_i\right) = k^2.$$

Now define $s_k := a_k + b_k + c_k$. Then we get

$$s_k = k^2 + 6k + 9 = (k+3)^2$$

as the total count of elements in all classes F_i for each k and therefore for each class C_{2k-4}^k the number of elements claimed in Table A.2 on page 131.

For the second diagonal and thus the classes C_{2k-6}^k we get a similar scheme:

k	$\#F_0$	$\#F_1$	$\#F_2$	$\#F_3$	$\#F_4$	$\#F_5$	$\Sigma\,\#F_i$	$\binom{k-1}{4}^2$
4	4	5	0	0	0	0	9	9
5	1	25	10	0	0	0	36	36
6	0	30	60	10	0	0	100	100
7	0	30	120	70	5	0	225	100
8	0	30	180	190	40	1	441	441
9	0	30	240	370	135	9	784	784

Between successive rows the operators are
$$\searrow\ \cdot\tfrac{5}{1}\ +\ \downarrow\ \cdot\tfrac{4}{2}\ +\ \searrow\ \cdot\tfrac{3}{3}\ +\ \downarrow\ \cdot\tfrac{2}{4}\ +\ \searrow\ \cdot\tfrac{1}{5}\ +\ \downarrow$$

$$\cdots \tag{5.16}$$

See below for the cases of the third and fourth diagonal:

k	$\#F_0$	$\#F_1$	$\#F_2$	$\#F_3$	$\#F_4$	$\#F_5$	$\#F_6$	$\#F_7$	$\Sigma\,\#F_i$	$\binom{k-1}{6}^2$
5	9	7	0	0	0	0	0	0	16	16
6	9	70	21	0	0	0	0	0	100	100
7	1	133	231	35	0	0	0	0	400	400
8	0	140	630	420	35	0	0	0	1225	1225
9	0	140	1050	1470	455	21	0	0	3136	3136

Between successive rows the operators are
$$\searrow\ \cdot\tfrac{7}{1}\ +\ \downarrow\ \cdot\tfrac{6}{2}\ +\ \searrow\ \cdot\tfrac{5}{3}\ +\ \downarrow\ \cdot\tfrac{4}{4}\ +\ \searrow\ \cdot\tfrac{3}{5}\ +\ \downarrow\ \cdot\tfrac{2}{6}\ +\ \searrow\ \cdot\tfrac{1}{7}\ +\ \downarrow$$

$$\cdots$$

(5.17) appears to the right of the $k=7$ row.

k	$\#F_0$	$\#F_1$	$\#F_2$	$\#F_3$	$\#F_4$	$\#F_5$	$\#F_6$	$\#F_7$	$\#F_8$	$\#F_9$	$\Sigma \#F_i$	$\binom{k-1}{8}^2$
6	16	9	0	0	0	0	0	0	0	0	25	25
7	36	153	36	0	0	0	0	0	0	0	225	225
8	16	477	648	84	0	0	0	0	0	0	1225	1225
9	1	621	2556	1596	126	0	0	0	0	0	4900	4900
10	0	630	5040	7560	2520	126	0	0	0	0	15876	15876
11	0	630	7560	19320	13860	2646	84	0	0	0	44100	44100

Between each pair of rows the inheritance pattern is indicated by:

$$\frac{.9}{1} \;+\; \frac{.8}{2} \;+\; \frac{.7}{3} \;+\; \frac{.6}{4} \;+\; \frac{.5}{5} \;+\; \frac{.4}{6} \;+\; \frac{.3}{7} \;+\; \frac{.2}{8} \;+\; \frac{.1}{9} \;+$$
$$\searrow \quad \downarrow \quad \searrow \quad \downarrow \quad \searrow \quad \downarrow \quad \searrow \quad \downarrow \quad \searrow \quad \downarrow \quad \searrow \quad \downarrow \quad \searrow \quad \downarrow \quad \searrow \quad \downarrow \quad \searrow \quad \downarrow$$

$$\cdots$$

$$(5.18)$$

Note that there is always a $k_0 \in \mathbb{N}$ such that for all $k > k_0$ there exist no fatherless cycles in the class $\mathcal{C}^k_{2k-2i}$ when looking at the inheritance along the $(i-1)$-th diagonal of Table A.1 on page 130 and Table A.2 on page 131. This is the case as by construction there can be no fatherless cycles left of the central class $\mathcal{C}^k_{k-1}$ (for odd k) as no class left of the central vertical axis of the triangles can be invariant under multiplication with λ_k. More specifically, the series of numbers of fatherless cycles for the inheritance along the i-th diagonal seems to be given by the numbers of elements in the classes $\mathcal{C}^{i+1}_2, \ldots, \mathcal{C}^{i+1}_0$. Furthermore we have $k_0 = 2(i+1)$ along the i-th diagonal.

So, as already mentioned earlier, the key to be able to fully prove the numbers arising in the inheritance process is to know how many fatherless elements there are in each step with $k \leq k_0$ for each diagonal. Unfortunately this is an open problem as of the time being.

Knowing the series of numbers of fatherless orbits for each diagonal, the full proof of the cardinalities of the classes of difference cycles along the diagonals should be possible, akin to the method that was presented here for the second diagonal.

During the proof one would have to proceed successively from diagonal to diagonal as the results of the earlier class cardinalities have to be used for this method.

In order to prove Conjecture 5.13 on page 113 it would thus remain to first of all show that the orbit numbers of the triangulations M^{2k-2} are as conjectured and of full length. Lemma 5.8 on page 106 then tells us that the complexes have the expected Euler characteristic of

$$\chi(M^{2k-2}) = \begin{cases} 0 & \text{for even } k \\ 4 & \text{for odd } k \end{cases},$$

and thus the homology of $S^{k-1} \times S^{k-1}$ as they are k-Hamiltonian in β^{2k} (which follows from Lemma 5.9 on page 107).

It finally would remain to be shown that M^{2k-2} is a combinatorial manifold. It seems as methods based on shelling arguments could be of use here.

Appendix A

Classes of difference cycles of M^{2k-2}

This appendix contains tables that list properties of the triangulations M^{2k-2} presented in Chapter 5 on page 97. Table A.1 on the next page describes the classes of difference cycles obtained by the inheritance process along the diagonals (see also Section 5.3.4 on page 120), Table A.2 on page 131 shows the size of the classes of Table A.1. In Table A.3 on page 132, the difference cycles of the triangulations M^{2k-2} are listed for $k = 2, \ldots, 5$. Table A.4 on page 133 contains calculated ($k \leq 12$) and conjectured ($k > 12$) parameters of M^{2k-2}. For more details of the triangulations M^{2k-2} see Chapter 5 on page 97.

Table A.1: The classes of difference cycles of the triangulations of $M^{2k-2} \cong S^{k-1} \times S^{k-1}$ for $k = 2,\ldots,8$. The inheritance scheme is denoted by the arrows $\searrow$ on the right pointing out the diagonals.

	1	2	3	4	5	6	7	8	9	10	11	12	13	14	15
	$\searrow^{1}$		$\searrow^{3}$		$\searrow^{5}$		$\searrow^{7}$		$\searrow^{9}$		$\searrow^{11}$		$\searrow^{13}$		$\searrow^{15}$
$k = 1$:	C_0^1														
$k = 2$:		C_2^2	C_0^2												
$k = 3$:			C_4^3	C_2^3	C_0^3										
$k = 4$:				C_6^4	C_4^4	C_2^4	C_0^4								
$k = 5$:					C_8^5	C_6^5	C_4^5	C_2^5	C_0^5						
$k = 6$:						C_{10}^6	C_8^6	C_6^6	C_4^6	C_2^6	C_0^6				
$k = 7$:							C_{12}^7	C_{10}^7	C_8^7	C_6^7	C_4^7	C_2^7	C_0^7		
$k = 8$:								C_{14}^8	C_{12}^8	C_{10}^8	C_8^8	C_6^8	C_4^8	C_2^8	C_0^8

Table A.2: The size of the classes of difference cycles of Table A.1 on the facing page. The inheritance scheme is again denoted by the arrows on the right pointing out the diagonals.

$k=1$:								1							$\nearrow^1$
$k=2$:							1		1						$\nearrow^3$
$k=3$:						1		4		1					$\nearrow^5$
$k=4$:					1		9		9		1				$\nearrow^7$
$k=5$:				1		16		36		16		1			$\nearrow^9$
$k=6$:			1		25		100		100		25		1		$\nearrow^{11}$
$k=7$:		1		36		225		400		225		36		1	$\nearrow^{13}$
$k=8$:	1		49		441		1225		1225		441		49		1 $\nearrow^{15}$

Table A.3: Difference cycles of the triangulations M^{2k-2} for $k = 2, \ldots, 5$. Difference cycles that are invariant under the multiplication $\cdot(2k-1)$ in $\mathbb{Z}_{4k}$ are marked with the superscript #, the ones invariant under the multiplication $\cdot(-1)$ in $\mathbb{Z}_{4k}$ are marked with the superscript *.

k	difference cycles
2	$(1:1:6)^*$, $(2:3:3)^*$.
3	$(1:1:1:8)^*$, $(1:1:2:5:3)^\#$, $(1:1:3:4:3)^*$, $(1:1:3:5:2)^\#$, $(1:2:1:4:4)^*$, $(2:2:3:2:3)^*$.
4	$(1:1:1:1:10)^*$, $(1:1:1:2:7:3)$, $(1:1:1:3:6:3)^*$, $(1:1:1:3:7:2)$, $(1:1:1:2:1:6:4)$, $(1:1:1:4:6:2)$, $(1:1:2:1:5:5)^*$, $(1:1:2:5:2:3)$, $(1:1:3:4:2:3)$, $(1:1:2:3:4:3:2)^*$, $(1:1:3:1:1:5:4)$, $(1:1:3:2:4:3:2)$, $(1:1:1:2:6:2:4)$, $(1:1:4:1:6:1:4)^*$, $(1:1:2:4:5:3:2)$, $(1:1:3:3:5:3:1:2)$, $(1:1:1:2:1:6:2:4)$, $(1:1:1:3:5:4:2)$, $(1:1:1:3:6:2:2)$, $(1:1:4:1:6:1:4)^*$, $(1:1:4:6:1:1:4)$, $(1:1:4:2:6:1:2:2)$, $(1:1:1:4:1:1:6:4)$, $(1:1:2:1:5:4:3)^\#$, $(1:1:2:1:5:3:4)$, $(1:1:2:1:2:5:3:4)$, $(1:1:2:1:1:5:3:4)$, $(1:1:2:1:5:4:3)^*$, $(1:2:1:2:5:4:3)^*$, $(1:1:2:2:5:2:3)$, $(1:1:2:3:4:2:3:2)$, $(1:1:2:3:4:3:2)^*$, $(1:1:2:3:4:2:1:4)$, $(1:1:2:4:2:1:4)$, $(1:1:3:1:2:5:3:3)$, $(1:1:2:4:5:1:3:3)$, $(1:1:2:4:5:2:1:3)$, $(1:1:3:1:4:4:3)^\#$, $(1:1:3:1:2:1:5:3:3)$, $(1:1:3:2:1:1:3)$, $(1:1:3:2:4:3:2)$, $(1:1:3:2:4:2:3)$, $(1:1:3:2:4:3:2)^*$, $(1:1:3:2:5:2:2)$, $(1:1:3:3:1:4:3)^*$, $(1:1:3:3:1:2)$, $(1:1:4:1:6:1:4)^*$, $(1:1:4:1:1:3:1:6:4)$, $(1:1:4:3:5:1:2)$, $(1:2:1:2:4:3:3)$, $(1:2:2:4:3:2:3)$, $(1:2:1:3:2:5:2:2)$, $(1:2:2:3:4:1:3)$, $(1:3:3:4:1:1:3)^*$, $(1:2:2:4:3:2:3)$, $(1:2:1:3:4:3:3)^\#$, $(1:2:3:4:2:2)$, $(1:2:2:4:2:2)$, $(1:2:1:4:3:5:1:2)$, $(1:2:1:2:4:2:3)$, $(1:2:3:4:2:2)$, $(1:2:1:3:3:3:2)$, $(1:2:3:4:2:2)$, $(1:2:1:3:3:3:3)^*$, $(1:2:1:3:3:3:2)$, $(1:2:1:4:3:3:2)$, $(1:2:1:2:4:2:3:3)$, $(1:2:2:3:4:2:3)$, $(2:2:2:3:2:2:3)$.

Table A.4: Calculated parameters of the conjectured series of centrally symmetric triangulations of $M^{2k-2} \cong S^{k-1} \times S^{k-1}$ for $k \leq 11$ and conjectured values (marked with $\star$) for higher values of k.

k	n	simplices	#facets in lk(0)	#diff.cycles	con. type	Aut	#Aut	χ
2	8	$8\binom{2}{1}=16$	$\frac{16\cdot3}{8}=6$	2	$S^1\times S^1$	$(D_8\times C_2)\rtimes C_2$	32	0
3	12	$12\binom{4}{2}=72$	$\frac{72\cdot5}{12}=30$	6	$S^2\times S^2$	$D_8\times S_3$	48	4
4	16	$16\binom{6}{3}=320$	$\frac{320\cdot7}{16}=140$	20	$S^3\times S^3$	$(C_2\times D_{16})\rtimes C_2$	64	0
5	20	$20\binom{8}{4}=1400$	$\frac{1400\cdot9}{20}=630$	70	$S^4\times S^4$	$D_8\times D_{10}$	80	4
6	24	$24\binom{10}{5}=6048$	$\frac{6048\cdot11}{24}=2772$	252	$S^5\times S^5$	$(C_3\times(C_8\rtimes C_2))\rtimes C_2$	96	0
7	28	$28\binom{12}{6}=25872$	$\frac{25872\cdot13}{28}=12012$	924	$S^6\times S^6$	$D_{14}\times D_8$	112	4
8	32	$32\binom{14}{7}=109824$	$\frac{109824\cdot15}{32}=51480$	3432	$S^7\times S^7$	$(C_2\times D_{32})\rtimes C_2$	128	0
9	36	$36\binom{16}{8}=463320$	$\frac{463320\cdot17}{36}=218790$	12870	$S^8\times S^8$	$D_{18}\times D_8$	144	4
10	40	$40\binom{18}{9}=1944800$	$\frac{1944800\cdot19}{40}=923780$	48620	$S^9\times S^9$	$(C_5\times(C_8\rtimes C_2))\rtimes C_2$	160	0
11	44	$40\binom{20}{10}=8129264$	$\frac{8129264\cdot21}{44}=3879876$	184756	$S^{10}\times S^{10}$	$D_{22}\times D_8$	176	4
$\vdots$	$\vdots$	$\vdots$	$\vdots$	$\vdots$	$\vdots$	$\vdots$	$\vdots$	$\vdots$
$k=2l$	$4k$	$4k\binom{2k-2}{k-1}^\star$	$(2k-1)\binom{2k-2}{k-1}^\star$	$\binom{2k-2}{k-1}^\star$	$S^{k-1}\times S^{k-1\,\star}$	?	$16k^\star$	$0^\star$
$k=2l+1$	$4k$	$4k\binom{2k-2}{k-1}^\star$	$(2k-1)\binom{2k-2}{k-1}^\star$	$\binom{2k-2}{k-1}^\star$	$S^{k-1}\times S^{k-1\,\star}$	?	$16k^\star$	$4^\star$

Appendix B

Facet lists of triangulations

B.1 A centrally symmetric 16-vertex triangulation of $(S^2 \times S^2)^{\#7}$

Given below is the list of the 224 facets of the triangulation of $(S^2 \times S^2)^{\#7}$ presented in Theorem 4.4 on page 82.

⟨1 3 5 7 9⟩, ⟨1 3 5 7 15⟩, ⟨1 3 5 8 13⟩, ⟨1 3 5 8 15⟩, ⟨1 3 5 9 13⟩, ⟨1 3 6 8 10⟩,
⟨1 3 6 8 12⟩, ⟨1 3 6 9 12⟩, ⟨1 3 6 9 16⟩, ⟨1 3 6 10 16⟩, ⟨1 3 7 9 15⟩, ⟨1 3 8 10 16⟩,
⟨1 3 8 11 14⟩, ⟨1 3 8 11 16⟩, ⟨1 3 8 12 13⟩, ⟨1 3 8 14 15⟩, ⟨1 3 9 11 14⟩, ⟨1 3 9 11 16⟩,
⟨1 3 9 12 13⟩, ⟨1 3 9 14 15⟩, ⟨1 4 5 9 12⟩, ⟨1 4 5 9 13⟩, ⟨1 4 5 11 13⟩, ⟨1 4 5 11 16⟩,
⟨1 4 5 12 10⟩, ⟨1 4 6 8 12⟩, ⟨1 4 6 8 13⟩, ⟨1 4 6 12 14⟩, ⟨1 4 6 13 15⟩, ⟨1 4 6 14 15⟩,
⟨1 4 7 10 12⟩, ⟨1 4 7 10 13⟩, ⟨1 4 7 12 15⟩, ⟨1 4 7 13 15⟩, ⟨1 4 8 9 12⟩, ⟨1 4 8 9 13⟩,
⟨1 4 10 12 16⟩, ⟨1 4 10 13 16⟩, ⟨1 4 11 13 16⟩,⟨1 4 12 14 15⟩, ⟨1 5 7 9 12⟩, ⟨1 5 7 10 12⟩,
⟨1 5 7 10 15⟩, ⟨1 5 8 11 13⟩, ⟨1 5 8 11 14⟩, ⟨1 5 8 14 15⟩, ⟨1 5 10 12 16⟩, ⟨1 5 10 14 15⟩,
⟨1 5 10 14 16⟩, ⟨1 5 11 14 16⟩, ⟨1 6 7 10 13⟩, ⟨1 6 7 10 16⟩, ⟨1 6 7 11 15⟩, ⟨1 6 7 11 16⟩,
⟨1 6 7 13 15⟩, ⟨1 6 8 10 13⟩, ⟨1 6 9 11 14⟩, ⟨1 6 9 11 16⟩, ⟨1 6 9 12 14⟩, ⟨1 6 11 14 15⟩,
⟨1 7 9 12 15⟩, ⟨1 7 10 11 14⟩, ⟨1 7 10 11 15⟩,⟨1 7 10 14 16⟩, ⟨1 7 11 14 16⟩, ⟨1 8 9 12 13⟩,
⟨1 8 10 13 16⟩, ⟨1 8 11 13 16⟩, ⟨1 9 12 14 15⟩,⟨1 10 11 14 15⟩,⟨2 3 5 7 11⟩, ⟨2 3 5 7 14⟩,
⟨2 3 5 11 13⟩, ⟨2 3 5 13 16⟩, ⟨2 3 5 14 16⟩, ⟨2 3 6 10 11⟩, ⟨2 3 6 10 14⟩, ⟨2 3 6 11 15⟩,
⟨2 3 6 12 14⟩, ⟨2 3 6 12 15⟩, ⟨2 3 7 10 11⟩, ⟨2 3 7 10 14⟩, ⟨2 3 8 9 11⟩, ⟨2 3 8 9 14⟩,
⟨2 3 8 11 16⟩, ⟨2 3 8 14 16⟩, ⟨2 3 9 11 15⟩, ⟨2 3 9 14 15⟩, ⟨2 3 11 13 16⟩, ⟨2 3 12 14 15⟩,
⟨2 4 5 7 9⟩, ⟨2 4 5 7 11⟩, ⟨2 4 5 9 15⟩, ⟨2 4 5 10 11⟩, ⟨2 4 5 10 15⟩, ⟨2 4 6 7 14⟩,
⟨2 4 6 7 16⟩, ⟨2 4 6 8 10⟩, ⟨2 4 6 8 16⟩, ⟨2 4 6 10 14⟩, ⟨2 4 7 9 15⟩, ⟨2 4 7 11 14⟩,
⟨2 4 7 12 13⟩, ⟨2 4 7 12 15⟩, ⟨2 4 7 13 16⟩, ⟨2 4 8 10 16⟩, ⟨2 4 10 11 14⟩, ⟨2 4 10 12 13⟩,
⟨2 4 10 12 15⟩, ⟨2 4 10 13 16⟩, ⟨2 5 7 9 14⟩, ⟨2 5 8 9 14⟩, ⟨2 5 8 9 15⟩, ⟨2 5 8 12 15⟩,
⟨2 5 8 12 16⟩, ⟨2 5 8 14 16⟩, ⟨2 5 10 11 13⟩,⟨2 5 10 12 13⟩, ⟨2 5 10 12 15⟩, ⟨2 5 12 13 16⟩,
⟨2 6 7 12 13⟩, ⟨2 6 7 12 14⟩, ⟨2 6 7 13 16⟩, ⟨2 6 8 9 11⟩, ⟨2 6 8 9 16⟩, ⟨2 6 8 10 11⟩,
⟨2 6 9 11 15⟩, ⟨2 6 9 13 15⟩, ⟨2 6 9 13 16⟩, ⟨2 6 12 13 15⟩, ⟨2 7 9 14 15⟩, ⟨2 7 10 11 14⟩,
⟨2 7 12 14 15⟩, ⟨2 8 9 12 13⟩, ⟨2 8 9 12 16⟩, ⟨2 8 9 13 15⟩, ⟨2 8 10 11 16⟩, ⟨2 8 12 13 15⟩,

⟨2 9 12 13 16⟩, ⟨2 10 11 13 16⟩, ⟨3 5 7 9 11⟩, ⟨3 5 7 10 12⟩, ⟨3 5 7 10 15⟩, ⟨3 5 7 12 16⟩,
⟨3 5 7 14 16⟩, ⟨3 5 8 13 15⟩, ⟨3 5 9 11 13⟩, ⟨3 5 10 12 13⟩, ⟨3 5 10 13 15⟩, ⟨3 5 12 13 16⟩,
⟨3 6 7 10 13⟩, ⟨3 6 7 10 16⟩, ⟨3 6 7 12 13⟩, ⟨3 6 7 12 16⟩, ⟨3 6 8 10 14⟩, ⟨3 6 8 12 14⟩,
⟨3 6 9 12 16⟩, ⟨3 6 10 11 15⟩, ⟨3 6 10 13 15⟩, ⟨3 6 12 13 15⟩, ⟨3 7 9 11 15⟩, ⟨3 7 10 11 15⟩,
⟨3 7 10 12 13⟩, ⟨3 7 10 14 16⟩, ⟨3 8 9 11 14⟩, ⟨3 8 10 14 16⟩, ⟨3 8 12 13 15⟩, ⟨3 8 12 14 15⟩,
⟨3 9 11 13 16⟩, ⟨3 9 12 13 16⟩, ⟨4 5 7 9 13⟩, ⟨4 5 7 11 13⟩, ⟨4 5 8 9 14⟩, ⟨4 5 8 9 15⟩,
⟨4 5 8 11 14⟩, ⟨4 5 8 11 15⟩, ⟨4 5 9 12 16⟩, ⟨4 5 9 14 16⟩, ⟨4 5 10 11 15⟩, ⟨4 5 11 14 16⟩,
⟨4 6 7 14 16⟩, ⟨4 6 8 9 11⟩, ⟨4 6 8 9 16⟩, ⟨4 6 8 10 12⟩, ⟨4 6 8 11 15⟩, ⟨4 6 8 13 15⟩,
⟨4 6 9 11 14⟩, ⟨4 6 9 14 16⟩, ⟨4 6 10 12 14⟩, ⟨4 6 11 14 15⟩, ⟨4 7 9 13 15⟩, ⟨4 7 10 12 13⟩,
⟨4 7 11 13 16⟩, ⟨4 7 11 14 16⟩, ⟨4 8 9 11 14⟩, ⟨4 8 9 12 16⟩, ⟨4 8 9 13 15⟩, ⟨4 8 10 12 16⟩,
⟨4 10 11 14 15⟩, ⟨4 10 12 14 15⟩, ⟨5 7 9 11 13⟩, ⟨5 7 9 12 16⟩, ⟨5 7 9 14 16⟩, ⟨5 8 10 12 14⟩,
⟨5 8 10 12 16⟩, ⟨5 8 10 14 16⟩, ⟨5 8 11 13 15⟩, ⟨5 8 12 14 15⟩, ⟨5 10 11 13 15⟩, ⟨5 10 12 14 15⟩,
⟨6 7 9 11 13⟩, ⟨6 7 9 11 15⟩, ⟨6 7 9 13 15⟩, ⟨6 7 11 13 16⟩, ⟨6 7 12 14 16⟩, ⟨6 8 10 11 15⟩,
⟨6 8 10 12 14⟩, ⟨6 8 10 13 15⟩, ⟨6 9 11 13 16⟩, ⟨6 9 12 14 16⟩, ⟨7 9 12 14 15⟩, ⟨7 9 12 14 16⟩,
⟨8 10 11 13 15⟩, ⟨8 10 11 13 16⟩.

B.2 A centrally symmetric 16-vertex triangulation of $S^4 \times S^2$

See below for a list of the 240 facets of M^6_{16} from Theorem 4.9 on page 93.

⟨1 2 3 4 7 12 14⟩,	⟨1 2 3 4 7 12 16⟩,	⟨1 2 3 4 7 13 14⟩,	⟨1 2 3 4 7 13 16⟩,	⟨1 2 3 4 9 12 14⟩,
⟨1 2 3 4 9 12 16⟩,	⟨1 2 3 4 9 14 16⟩,	⟨1 2 3 4 13 14 16⟩,	⟨1 2 3 6 7 12 14⟩,	⟨1 2 3 6 7 12 16⟩,
⟨1 2 3 6 7 13 14⟩,	⟨1 2 3 6 7 13 16⟩,	⟨1 2 3 6 9 10 12⟩,	⟨1 2 3 6 9 10 13⟩,	⟨1 2 3 6 9 12 16⟩,
⟨1 2 3 6 9 13 16⟩,	⟨1 2 3 6 10 11 12⟩,	⟨1 2 3 6 10 11 13⟩,	⟨1 2 3 6 11 12 14⟩,	⟨1 2 3 6 11 13 14⟩,
⟨1 2 3 9 10 11 12⟩,	⟨1 2 3 9 10 11 13⟩,	⟨1 2 3 9 11 12 14⟩,	⟨1 2 3 9 11 13 14⟩,	⟨1 2 3 9 13 14 16⟩,
⟨1 2 4 7 12 14 15⟩,	⟨1 2 4 7 12 15 16⟩,	⟨1 2 4 7 13 14 15⟩,	⟨1 2 4 7 13 15 16⟩,	⟨1 2 4 9 12 14 16⟩,
⟨1 2 4 12 14 15 16⟩,	⟨1 2 4 13 14 15 16⟩,	⟨1 2 6 7 12 14 16⟩,	⟨1 2 6 7 13 14 15⟩,	⟨1 2 6 7 13 15 16⟩,
⟨1 2 6 7 14 15 16⟩,	⟨1 2 6 9 10 11 12⟩,	⟨1 2 6 9 10 11 13⟩,	⟨1 2 6 9 11 12 14⟩,	⟨1 2 6 9 11 13 15⟩,
⟨1 2 6 9 11 14 15⟩,	⟨1 2 6 9 12 14 16⟩,	⟨1 2 6 9 13 15 16⟩,	⟨1 2 6 9 14 15 16⟩,	⟨1 2 6 11 13 14 15⟩,
⟨1 2 7 12 14 15 16⟩,	⟨1 2 9 11 13 14 15⟩,	⟨1 2 9 13 14 15 16⟩,	⟨1 3 4 7 12 14 16⟩,	⟨1 3 4 7 13 14 16⟩,
⟨1 3 4 9 12 14 16⟩,	⟨1 3 6 7 12 14 16⟩,	⟨1 3 6 7 13 14 16⟩,	⟨1 3 6 8 9 10 11⟩,	⟨1 3 6 8 9 10 13⟩,
⟨1 3 6 8 9 11 14⟩,	⟨1 3 6 8 9 13 14⟩,	⟨1 3 6 8 10 11 13⟩,	⟨1 3 6 8 11 13 14⟩,	⟨1 3 6 9 10 11 12⟩,
⟨1 3 6 9 11 12 14⟩,	⟨1 3 6 9 12 14 16⟩,	⟨1 3 6 9 13 14 16⟩,	⟨1 3 8 9 10 11 13⟩,	⟨1 3 8 9 11 13 14⟩,
⟨1 4 7 8 10 11 13⟩,	⟨1 4 7 8 10 11 15⟩,	⟨1 4 7 8 10 13 16⟩,	⟨1 4 7 8 10 15 16⟩,	⟨1 4 7 8 11 13 15⟩,
⟨1 4 7 8 12 14 15⟩,	⟨1 4 7 8 12 14 16⟩,	⟨1 4 7 8 12 15 16⟩,	⟨1 4 7 8 13 14 15⟩,	⟨1 4 7 8 13 14 16⟩,
⟨1 4 7 10 11 13 15⟩,	⟨1 4 7 10 13 15 16⟩,	⟨1 4 8 10 11 13 15⟩,	⟨1 4 8 10 13 15 16⟩,	⟨1 4 8 12 14 15 16⟩,
⟨1 4 8 13 14 15 16⟩,	⟨1 6 7 8 10 11 13⟩,	⟨1 6 7 8 10 11 15⟩,	⟨1 6 7 8 10 13 16⟩,	⟨1 6 7 8 10 15 16⟩,
⟨1 6 7 8 11 13 15⟩,	⟨1 6 7 8 13 14 15⟩,	⟨1 6 7 8 13 14 16⟩,	⟨1 6 7 8 14 15 16⟩,	⟨1 6 7 10 11 13 15⟩,
⟨1 6 7 10 13 15 16⟩,	⟨1 6 8 9 10 11 15⟩,	⟨1 6 8 9 10 13 16⟩,	⟨1 6 8 9 10 15 16⟩,	⟨1 6 8 9 11 14 15⟩,
⟨1 6 8 9 13 14 16⟩,	⟨1 6 8 9 14 15 16⟩,	⟨1 6 8 11 13 14 15⟩,	⟨1 6 9 10 11 13 15⟩,	⟨1 6 9 10 13 15 16⟩,
⟨1 7 8 12 14 15 16⟩,	⟨1 8 9 10 11 13 15⟩,	⟨1 8 9 10 13 15 16⟩,	⟨1 8 9 11 13 14 15⟩,	⟨1 8 9 13 14 15 16⟩,

B.2. A CENTRALLY SYMMETRIC 16-VERTEX TRIANGULATION OF $S^4 \times S^2$

⟨2 3 4 5 7 10 11⟩,	⟨2 3 4 5 7 10 16⟩,	⟨2 3 4 5 7 11 14⟩,	⟨2 3 4 5 7 14 16⟩,	⟨2 3 4 5 9 10 11⟩,
⟨2 3 4 5 9 10 12⟩,	⟨2 3 4 5 9 11 14⟩,	⟨2 3 4 5 9 12 16⟩,	⟨2 3 4 5 9 14 16⟩,	⟨2 3 4 5 10 12 16⟩,
⟨2 3 4 7 10 11 12⟩,	⟨2 3 4 7 10 12 16⟩,	⟨2 3 4 7 11 12 14⟩,	⟨2 3 4 7 13 14 16⟩,	⟨2 3 4 9 10 11 12⟩,
⟨2 3 4 9 11 12 14⟩,	⟨2 3 5 6 9 10 12⟩,	⟨2 3 5 6 9 10 13⟩,	⟨2 3 5 6 9 11 13⟩,	⟨2 3 5 6 9 11 14⟩,
⟨2 3 5 6 9 12 16⟩,	⟨2 3 5 6 9 14 16⟩,	⟨2 3 5 6 10 11 12⟩,	⟨2 3 5 6 10 11 13⟩,	⟨2 3 5 6 11 12 14⟩,
⟨2 3 5 6 12 14 16⟩,	⟨2 3 5 7 10 11 12⟩,	⟨2 3 5 7 10 12 16⟩,	⟨2 3 5 7 11 12 14⟩,	⟨2 3 5 7 12 14 16⟩,
⟨2 3 5 9 10 11 13⟩,	⟨2 3 6 7 12 14 16⟩,	⟨2 3 6 7 13 14 16⟩,	⟨2 3 6 9 11 13 14⟩,	⟨2 3 6 9 13 14 16⟩,
⟨2 4 5 7 10 11 12⟩,	⟨2 4 5 7 10 12 15⟩,	⟨2 4 5 7 10 15 16⟩,	⟨2 4 5 7 11 12 14⟩,	⟨2 4 5 7 12 14 15⟩,
⟨2 4 5 7 14 15 16⟩,	⟨2 4 5 9 10 11 12⟩,	⟨2 4 5 9 11 12 14⟩,	⟨2 4 5 9 12 14 16⟩,	⟨2 4 5 10 12 15 16⟩,
⟨2 4 5 12 14 15 16⟩,	⟨2 4 7 10 12 15 16⟩,	⟨2 4 7 13 14 15 16⟩,	⟨2 5 6 9 10 11 12⟩,	⟨2 5 6 9 10 11 13⟩,
⟨2 5 6 9 11 12 14⟩,	⟨2 5 6 9 12 14 16⟩,	⟨2 5 7 10 12 15 16⟩,	⟨2 5 7 12 14 15 16⟩,	⟨2 6 7 13 14 15 16⟩,
⟨2 6 9 11 13 14 15⟩,	⟨2 6 9 13 14 15 16⟩,	⟨3 4 5 7 8 10 11⟩,	⟨3 4 5 7 8 10 16⟩,	⟨3 4 5 7 8 11 12⟩,
⟨3 4 5 7 8 12 16⟩,	⟨3 4 5 7 11 12 14⟩,	⟨3 4 5 7 12 14 16⟩,	⟨3 4 5 8 9 10 11⟩,	⟨3 4 5 8 9 10 12⟩,
⟨3 4 5 8 9 11 12⟩,	⟨3 4 5 8 10 12 16⟩,	⟨3 4 5 9 11 12 14⟩,	⟨3 4 5 9 12 14 16⟩,	⟨3 4 7 8 10 11 12⟩,
⟨3 4 7 8 10 12 16⟩,	⟨3 4 8 9 10 11 12⟩,	⟨3 5 6 8 9 10 12⟩,	⟨3 5 6 8 9 10 13⟩,	⟨3 5 6 8 9 11 12⟩,
⟨3 5 6 8 9 11 13⟩,	⟨3 5 6 8 10 11 12⟩,	⟨3 5 6 8 10 11 13⟩,	⟨3 5 6 9 11 12 14⟩,	⟨3 5 6 9 12 14 16⟩,
⟨3 5 7 8 10 11 12⟩,	⟨3 5 7 8 10 12 16⟩,	⟨3 5 8 9 10 11 13⟩,	⟨3 6 8 9 10 11 12⟩,	⟨3 6 8 9 11 13 14⟩,
⟨4 5 7 8 10 11 13⟩,	⟨4 5 7 8 10 13 16⟩,	⟨4 5 7 8 11 12 15⟩,	⟨4 5 7 8 11 13 15⟩,	⟨4 5 7 8 12 14 15⟩,
⟨4 5 7 8 12 14 16⟩,	⟨4 5 7 8 13 14 15⟩,	⟨4 5 7 8 13 14 16⟩,	⟨4 5 7 10 11 12 15⟩,	⟨4 5 7 10 11 13 15⟩,
⟨4 5 7 10 13 15 16⟩,	⟨4 5 7 13 14 15 16⟩,	⟨4 5 8 9 10 11 13⟩,	⟨4 5 8 9 10 12 15⟩,	⟨4 5 8 9 10 13 15⟩,
⟨4 5 8 9 11 12 15⟩,	⟨4 5 8 9 11 13 15⟩,	⟨4 5 8 10 12 15 16⟩,	⟨4 5 8 10 13 15 16⟩,	⟨4 5 8 12 14 15 16⟩,
⟨4 5 8 13 14 15 16⟩,	⟨4 5 9 10 11 12 15⟩,	⟨4 5 9 10 11 13 15⟩,	⟨4 7 8 10 12 15 16⟩,	⟨4 7 8 10 12 15 16⟩,
⟨4 8 9 10 11 12 15⟩,	⟨4 8 9 10 11 13 15⟩,	⟨5 6 7 8 10 11 13⟩,	⟨5 6 7 8 10 11 15⟩,	⟨5 6 7 8 10 13 15⟩,
⟨5 6 7 8 11 13 15⟩,	⟨5 6 7 10 11 13 15⟩,	⟨5 6 8 9 10 12 15⟩,	⟨5 6 8 9 10 13 15⟩,	⟨5 6 8 9 11 12 15⟩,
⟨5 6 8 9 11 13 15⟩,	⟨5 6 8 10 11 12 15⟩,	⟨5 6 9 10 11 12 15⟩,	⟨5 6 9 10 11 13 15⟩,	⟨5 7 8 10 11 12 15⟩,
⟨5 7 8 10 12 15 16⟩,	⟨5 7 8 10 13 15 16⟩,	⟨5 7 8 12 14 15 16⟩,	⟨5 7 8 13 14 15 16⟩,	⟨6 7 8 10 13 15 16⟩,
⟨6 7 8 13 14 15 16⟩,	⟨6 8 9 10 11 12 15⟩,	⟨6 8 9 10 13 15 16⟩,	⟨6 8 9 11 13 14 15⟩,	⟨6 8 9 13 14 15 16⟩.

Appendix C

The GAP package simpcomp

simpcomp[1] [44, 45] is an extension (a so-called *package*) to GAP [51], the well known system for computational discrete algebra. In contrast to the package homology [38] which focuses on simplicial homology computation, simpcomp claims to provide the user with a broader spectrum of functionality regarding simplicial constructions. simpcomp allows the user to interactively construct (abstract) simplicial complexes and to compute their properties in the GAP shell. The package caches computed properties of a simplicial complex, thus avoiding unnecessary computations, internally handles the vertex labeling of the complexes and insures the consistency of a simplicial complex throughout all operations. Furthermore, it makes use of GAP's expertise in groups and group operations. For example, automorphism groups and fundamental groups of complexes can be computed and examined further within the GAP system.

As of the time being, simpcomp relies on the GAP package homology [38] for its homology computation, but also provides the user with an own (co-)homology algorithm in case the package homology is not available. For automorphism group computation the GAP package GRAPE [121] is used, which in turn uses the program nauty by Brendan McKay [94]. An internal automorphism group calculation algorithm in used as fallback if the GRAPE package is not available.

The package includes an extensive manual in which all functionality of simpcomp is documented, see [44].

---

[1]The software simpcomp presented in this chapter was developed together with Jonathan Spreer. All what is presented in this chapter is joint work and effort.

C.1   What is new

simpcomp allows the user to interactively construct complexes and to compute their properties in the GAP shell. Furthermore, it makes use of GAP's expertise in groups and group operations. For example, automorphism groups and fundamental groups of complexes can be computed and examined further within the GAP system. Apart from supplying a facet list, the user can as well construct simplicial complexes from a set of generators and a prescribed automorphism group – the latter form being the common in which a complex is presented in a publication. This feature is to our knowledge unique to simpcomp. Furthermore, simpcomp as of Version 1.3.0 supports the construction of simplicial complexes of prescribed dimension, vertex number and transitive automorphism group as described in [90], [29].

As of version 1.4.0, simpcomp supports *simplicial blowups*, i.e. the resolutions of ordinary double points in combinatorial 4-pseudomanifolds. This functionality is to the author's knowledge not provided by any other software package so far.

Furthermore, simpcomp has an extensive library of known triangulations of manifolds. This is the first time that they are easily accessible without having to look them up in the literature [84], [29], or online [89]. This allows the user to work with many different known triangulations without having to construct them first. As of version 1.3.0, the library contains triangulations of roughly 650 manifolds and roughly 7000 pseudomanifolds, including all vertex transitive triangulations from [89]. Most properties that simpcomp can handle are precomputed for complexes in the library. Searching in the library is possible by the complexes' names as well as some of their properties (such as f-, g- and h-vectors and their homology).

C.2   simpcomp benefits

simpcomp is written entirely in the GAP scripting language, thus giving the user the possibility to see behind the scenes and to customize or alter simpcomp functions if needed.

The main benefit when working with simpcomp over implementing the needed functions from scratch is that simpcomp encapsulates all methods and properties of a simplicial complex in a new GAP object type (as an abstract data type). This way, among other things, simpcomp can transparently cache properties already calculated, thus preventing unnecessary double calculations. It also takes care of the error-prone vertex labeling of a complex.

simpcomp provides the user with functions to save and load the simplicial complexes to and from files and to import and export a complex in various formats (e.g. from and to polymake/TOPAZ [52], Macaulay2 [55], LATEX, etc.).

In contrast to the software package polymake [52] providing the most efficient algorithms for each task in form of a heterogeneous package (where algorithms are implemented in various languages), the primary goal when developing simpcomp was not efficiency (this is already limited by the GAP scripting language), but rather ease of use and ease of extensibility by the user in the GAP language with all its mathematical and algebraic capabilities.

The package includes an extensive manual (see [44]) in which all functionality of simpcomp is documented and also makes use of GAP's built in help system so that all the documentation is available directly from the GAP prompt in an interactive way.

C.3 Some operations and constructions that simpcomp supports

simpcomp implements many standard and often needed functions for working with simplicial complexes. These functions can be roughly divided into three groups: (i) functions generating simplicial complexes (ii) functions to construct new complexes from old and (iii) functions calculating properties of complexes – for a full list of supported features see the documentation [44].

simpcomp furthermore implements a variety of functions connected to *bistellar moves* (also known as *Pachner moves* [111], see Section 1.5 on page 24) on simplicial complexes. For example, simpcomp can be used to construct randomized spheres or randomize a given complex. Another prominent application of bistellar moves implemented in simpcomp is a heuristic algorithm that determines whether a simplicial complex is a *combinatorial manifold* (i.e. that each link is PL homeomorphic to the boundary of the simplex). This algorithm was first presented by Lutz and Anders Björner [22]. It uses a simulated annealing type strategy in order to minimize vertex numbers of triangulations while leaving the PL homeomorphism type invariant.

The package also supports *slicings* of 3-manifolds (known as discrete *normal surfaces*, see [70], [58], [127]) and related constructions as well as functions related to polyhedral Morse theory.

The first group contains functions that create a simplicial complex object from a facet list (`SCFromFacets`), from a group operation on some generating simplices (`SCFromGenerators`) and from difference cycles (`SCFromDifferenceCycles`). Another way to obtain known (in some cases minimal) triangulations of manifolds is to use the simplicial complex library, see Section C.4 on the facing page. Also in this group are functions that generate some standard (and often needed) triangulations, e.g. that of the boundary of the n-simplex (`SCBdSimplex`), the n-cross polytope (`SCBdCrossPolytope`) and the empty complex (`SCEmpty`).

The second group contains functions that take one or more simplicial complexes as their arguments and return a new simplicial complex. Among these are the functions to compute links and stars of faces (`SCLink`, `SCStar`), to form a connected sum (`SCConnectedSum`), a cartesian product (`SCCartesianProduct`), a join (`SCJoin`) or a suspension (`SCSuspension`) of (a) simplicial complexe(s).

The third and by far the largest group is that of the functions computing properties of simplicial complexes. Just to name a few, `simpcomp` can compute the f-, g- and h-vector of a complex (`SCFVector`, `SCGVector`, `SCHVector`), its Euler characteristic (`SCEulerCharacteristic`), the face lattice and skeletons of different dimensions (`SCFaceLattice`, `SCFaces`), the automorphism group of a complex (`SCAutomorphismGroup`), homology and cohomology with explicit bases (`SCHomology`, `SCCohomology`, `SCHomologyBasis`, `SCHomologyBasisAsSimplices`, `SCCohomologyBasis`, `SCCohomologyBasisAsSimplices`), the cup product (`SCCupProduct`), the intersection form for closed, oriented 4-manifolds (`SCIntersectionForm`), spanning trees (`SCSpanningTree`), fundamental groups (`SCFundamentalGr`-

oup), dual graphs (SCDualGraph), connected and strongly connected components (SCConnectedComponents, SCStronglyConnectedComponents).

simpcomp can furthermore determine whether two simplicial complexes are combinatorially isomorphic and contains a heuristic algorithm based on bistellar flips (cf. [89, 90]) that tries to determine whether two simplicial complexes are PL homeomorphic.

C.4 The simplicial complex library of simpcomp

simpcomp contains a library of simplicial complexes on few vertices, most of them (combinatorial) triangulations of manifolds and pseudomanifolds. The user can load these known triangulations from the library in order to study their properties or to construct new triangulations out of the known ones. For example, a user could try to determine the topological type of a given triangulation – which can be quite tedious if done by hand – by establishing a PL equivalence to a complex in the library.

Among other known triangulations, the library contains all of the vertex transitive triangulations of d-manifolds, $d \leq 11$ with few ($n \leq 13$ and $n \leq 15$ for $d = 2, 3, 9, 10, 11$) vertices classified by Frank Lutz that can be found on his "Manifold Page" [89], along with some triangulations of sphere bundles and vertex transitive triangulations of pseudomanifolds.

C.5 Demonstration sessions with simpcomp

This section contains a small demonstration of the capabilities of simpcomp in form of two demonstration sessions.

C.5.1 First demonstration session

M. Casella and W. Kühnel constructed a triangulated K3 surface with the minimum number of 16 vertices in [29]. They presented it in terms of the complex obtained by the automorphism group $G \cong AGL(1, \mathbb{F}_{16})$ given by the five generators

$$G = \left\{ \begin{array}{l} (1\,2)(3\,4)(5\,6)(7\,8)(9\,10)(11\,12)(13\,14)(15\,16), \\ (1\,3)(2\,4)(5\,7)(6\,8)(9\,11)(10\,12)(13\,15)(14\,16), \\ (1\,5)(2\,6)(3\,7)(4\,8)(9\,13)(10\,14)(11\,15)(12\,16), \\ (1\,9)(2\,10)(3\,11)(4\,12)(5\,13)(6\,14)(7\,15)(8\,16), \\ (2\,13\,15\,11\,14\,3\,5\,8\,16\,7\,4\,9\,10\,6\,12) \end{array} \right\}$$

acting on the two generating simplices $\Delta_1 = \langle 2, 3, 4, 5, 9 \rangle$ and $\Delta_2 = \langle 2, 5, 7, 10, 11 \rangle$.
It turned out to be a non-trivial problem to show (i) that the complex obtained
is a combinatorial 4-manifold and (ii) to show that it is homeomorphic to the $K3$
surface as topological 4-manifold.

This turns out to be a rather easy task using simpcomp, as will be shown below.
We will fire up GAP, load simpcomp and then construct the complex from its
representation given above.

```
1  $ gap

         ########            ######        ##########              ###
       ############          ######        ###########            ####
5      #############        ########       #############          #####
      ###############       ########       #####  ######          #####
      ######      #        #########       #####   #####         ######
      ######                ##########     #####   #####        #######
      #####                 ##### ####      #####   ######      ########
10    ####                  #####  #####    ############      ###  ####
      #####     #######      ####    ####    ##########      ####  ####
      #####     #######      #####    #####   ######         ####  ####
      #####     #######      #####    #####   #####         ############
      #####       #####     ################   #####        ############
15    ######      #####     ################   #####        ############
      ###############       #################  #####            ####
      ##############        #####       #####  #####            ####
       #############        #####       #####  #####            ####
        ########            #####       #####  #####            ####
20
       Information  at:   http://www.gap-system.org
       Try '?help' for help. See also   '?copyright' and   '?authors'

       Loading  the  library.  Please  be  patient,  this  may  take  a  while.
25 GAP4,  Version:  4.4.12 of  17-Dec-2008,  i686-pc-linux-gnu-gcc
   Components:   small 2.1,  small2 2.0,  small3 2.0,  small4 1.0,  small5 1.0,
                 small6 1.0,  small7 1.0,  small8 1.0,  small9 1.0,  small10 0.2,
                 id2 3.0,  id3 2.1,  id4 1.0,  id5 1.0,  id6 1.0,  id9 1.0,  id10 0.1,
                 trans 1.0,  prim 2.1   loaded.
30 Packages:     AClib 1.1,  Polycyclic 1.1,  Alnuth 2.1.3,  CrystCat 1.1.2,
```

Cryst 4.1.4, AutPGrp 1.2, CRISP 1.2.1, CTblLib 1.1.3,
TomLib 1.1.2, FactInt 1.4.10, FGA 1.1.0.1, GAPDoc 0.9999,
Homology 1.4.2, IRREDSOL 1.0.9, LAGUNA 3.3.1, Sophus 1.21,
Polenta 1.2.1, ResClasses 2.1.1 loaded.

```
35 gap> LoadPackage("simpcomp");; #load the package
   Loading simpcomp 1.4.0
   by F. Effenberger and J. Spreer
   http://www.igt.uni-stuttgart.de/LstDiffgeo/simpcomp
   gap> SCInfoLevel(0);; #suppress simpcomp info messages
40 gap> G:=Group((1,2)(3,4)(5,6)(7,8)(9,10)(11,12)(13,14)(15,16),
   >  (1,3)(2,4)(5,7)(6,8)(9,11)(10,12)(13,15)(14,16),
   >  (1,5)(2,6)(3,7)(4,8)(9,13)(10,14)(11,15)(12,16),
   >  (1,9)(2,10)(3,11)(4,12)(5,13)(6,14)(7,15)(8,16),
   >  (2,13,15,11,14,3,5,8,16,7,4,9,10,6,12));;
45 gap> K3:=SCFromGenerators(G,[[2,3,4,5,9],[2,5,7,10,11]]);
   [SimplicialComplex

   Properties known: Dim, Facets, Generators, Name, VertexLabels.

50 Name="complex from generators under group ((C2 x C2 x C2 x C2) : C5) : C3"
   Dim=4

   /SimplicialComplex]
   gap> K3.F;
55 [ 16, 120, 560, 720, 288 ]
   gap> K3.Chi;
   24
   gap> K3.Homology;
```

We first compute the f-vector, the Euler characteristic and the homology groups of K3.

```
59 [ [ 0, [  ] ], [ 0, [  ] ], [ 22, [  ] ], [ 0, [  ] ], [ 1, [  ] ] ]
60 gap> K3.IsManifold;
   true
   gap> K3.IntersectionFormParity;
   0
   gap> K3.IntersectionFormSignature;
```

Now we verify that the complex K3 is a combinatorial manifold using the heuristic algorithm based on bistellar moves described above.

```
65 [ 22, 3, 19 ]
   gap> K3.FundamentalGroup;
```

In a next step we compute the parity and the signature of the intersection form of the complex K3.

```
67 <fp group with 105 generators>
   gap> Size(last);
   1
70 gap> K3.Neighborliness;
```

This means that the intersection form of the complex K3 is even. It has dimension 22 and signature $19 - 3 = 16$. Furthermore, K3 is simply connected as can either be verified by showing that the fundamental group is trivial or by checking that the complex is 3-neighborly.

```
71 3
   gap> SCInfoLevel(2);
   gap> K3.IsTight;
   #I  SCIsTight: complex is (k+1)-neighborly 2k-manifold and thus tight.
75 true
```

It now follows from a theorem of M. Freedman [49] that the complex is in fact homeomorphic to a K3 surface because it has the same (even) intersection form. Furthermore, K3 is a tight triangulation as it is a 3-neighborly triangulation of a 4-manifold, see Theorem 1.47 on page 30.

C.5.2   Second demonstration session

In this session the triangulation M_{15}^4 due to Bagchi and Datta [11] is constructed and checked to lie in $\mathcal{K}(4)$, see Chapter 3 on page 53.

In the listing below, first the 5-ball B_{30}^5 is constructed via its facet list (lines 39-60) after simpcomp was loaded (line 35-36). Then some properties of B_{30}^5 are calculated (lines 62-68). This is followed by the calculation of the boundary of B_{30}^5 (lines 72-80) and the process of adding three handles between three facet pairs (δ_i, δ_i'), $1 \leq i \leq 3$, cf. Figure 3.1 on page 64 (lines 82-114) to finally obtain M_{15}^4 (line 104). Subsequently, some properties of M_{15}^4 are calculated (lines 116-129), it is verified via bistellar moves that M_{15}^4 is a combinatorial manifold (line 131-158) and it is checked (also using bistellar moves) that $M_{15}^4 \in \mathcal{K}(4)$ (lines 160-188). In a last step, it is verified that M_{15}^4 is tight (lines 198-192) and the multiplicity vector of the PL Morse function given by $v_1 < v_2 < \cdots < v_{15}$ is computed (lines 193-208).

```
1 $ gap


         ########        ######        ##########        ###
        #############     ######        ###########       ####
5      ##############    ########       ############      #####
       ##############    ########     ##### ######        #####
       ######     #      #########    #####   #####       ######
       ######            ##########   #####   #####       #######
       #####             ##### ####   #####   ######      ########
10     ####              ##### #####  #############  ###  ####
       #####  #######    ####   ####  ##########     #### ####
       #####  #######    #####  ##### ######         #### ####
       #####  #######    #####  ##### #####          #############
       #####   #####    ###############  #####       #############
15     ######  #####    ###############  #####       #############
       ###############  ################# #####            ####
       #############    #####      #####  #####            ####
       ############     #####      #####  #####            ####
        ########        #####      ##### #####             ####
20
        Information at:  http://www.gap-system.org
        Try '?help' for help. See also  '?copyright' and  '?authors'

        Loading the library. Please be patient, this may take a while.
25 GAP4, Version: 4.4.12 of 17-Dec-2008, i686-pc-linux-gnu-gcc
   Components:  small 2.1, small2 2.0, small3 2.0, small4 1.0, small5 1.0,
                small6 1.0, small7 1.0, small8 1.0, small9 1.0, small10 0.2,
                id2 3.0, id3 2.1, id4 1.0, id5 1.0, id6 1.0, id9 1.0, id10 0.1,
                trans 1.0, prim 2.1 loaded.
30 Packages:    AClib 1.1, Polycyclic 1.1, Alnuth 2.1.3, CrystCat 1.1.2,
                Cryst 4.1.4, AutPGrp 1.2, CRISP 1.2.1, CTblLib 1.1.3,
                TomLib 1.1.2, FactInt 1.4.10, FGA 1.1.0.1, GAPDoc 0.9999,
                Homology 1.4.2, IRREDSOL 1.0.9, LAGUNA 3.3.1, Sophus 1.21,
                Polenta 1.2.1, ResClasses 2.1.1 loaded.

35 gap> LoadPackage("simpcomp");; #load the package
   Loading simpcomp 1.4.0
   by F. Effenberger and J. Spreer
   http://www.igt.uni-stuttgart.de/LstDiffgeo/simpcomp
   gap> facets:=
40 > [ [ 1, 2, 6, 7, 12, 11 ], [ 1, 2, 4, 6, 7, 12 ], [ 1, 2, 3, 4, 6, 7 ],
   >   [ 1, 2, 3, 4, 5, 6 ], [ 2, 3, 4, 5, 6, 30 ], [ 3, 4, 5, 6, 30, 29 ],
   >   [ 3, 4, 5, 28, 29, 30 ], [ 3, 5, 27, 28, 29, 30 ],
   >   [ 26, 27, 28, 29, 30, 3 ], [ 1, 2, 7, 11, 12, 14 ],
   >   [ 1, 2, 11, 12, 13, 14 ], [ 1, 11, 12, 13, 15, 14 ],
45 >   [ 1, 12, 13, 14, 15, 25 ], [ 1, 13, 14, 15, 24, 25 ],
   >   [ 13, 14, 15, 23, 24, 25 ], [ 13, 15, 22, 23, 24, 25 ],
   >   [ 21, 22, 23, 24, 25, 13 ], [ 2, 6, 7, 9, 12, 11 ], [ 6, 7, 8, 9, 11, 12 ],
   >   [ 6, 7, 8, 9, 10, 11 ], [ 20, 7, 8, 10, 9, 11 ], [ 19, 20, 8, 9, 10, 11 ],
```

```
     >    [ 18, 19, 20, 8, 10, 9 ], [ 17, 18, 19, 20, 8, 10 ],
50   >    [ 16, 17, 18, 19, 20, 8 ] ];;

     gap>  b5_30:=SCFromFacets(facets);
     [SimplicialComplex

55   Properties known: Dim, Facets, Name, VertexLabels.

        Name="unnamed complex 1"
        Dim=5

60   /SimplicialComplex]

     gap>  b5_30.F;
     [ 30, 135, 260, 255, 126, 25 ]

65   gap>  b5_30.Chi;
     1

     gap>  b5_30.Homology;
     [ [ 0, [  ] ], [ 0, [  ] ], [ 0, [  ] ], [ 0, [  ] ], [ 0, [  ] ],
70   [ 0, [  ] ] ]

     gap>  bd:=b5_30.Boundary;
     [SimplicialComplex

75   Properties known: Dim, Facets, Name, VertexLabels.

        Name="Bd(unnamed complex 1)"
        Dim=4

80   /SimplicialComplex]

     gap>  handle1:=bd.HandleAddition([1..5],[16..20]);
     [SimplicialComplex

85   Properties known: Dim, Facets, Name, VertexLabels.

        Name="Bd(unnamed complex 1) handle ([ 1, 2, 3, 4, 5 ]=[ 16, 17, 18, 19, 20 ])\
     "
        Dim=4
90
     /SimplicialComplex]

     gap>  handle2:=handle1.HandleAddition([6..10],[21..25]);
     [SimplicialComplex
95
     Properties known: Dim, Facets, Name, VertexLabels.
```

```
      Name="Bd(unnamed complex 1) handle ([ 1, 2, 3, 4, 5 ]=[ 16, 17, 18, 19, 20 ])\
      handle ([ 6, 7, 8, 9, 10 ]=[ 21, 22, 23, 24, 25 ])"
100   Dim=4

      /SimplicialComplex]

      gap> m4_15:=handle2.HandleAddition([11..15],[26..30]);
105   [SimplicialComplex

      Properties known: Dim, Facets, Name, VertexLabels.

      Name="Bd(unnamed complex 1) handle ([ 1, 2, 3, 4, 5 ]=[ 16, 17, 18, 19, 20 ])\
110   handle ([ 6, 7, 8, 9, 10 ]=[ 21, 22, 23, 24, 25 ]) handle ([ 11, 12, 13, 14, \
      15 ]=[ 26, 27, 28, 29, 30 ])"
      Dim=4

      /SimplicialComplex]
115
      gap> SCRename(m4_15,"M`4_15");
      true

      gap> m4_15.F;
120   [ 15, 105, 230, 240, 96 ]

      gap> m4_15.Chi;
      -4

125   gap> m4_15.Homology;
      [ [ 0, [  ] ], [ 3, [  ] ], [ 0, [  ] ], [ 2, [ 2 ] ], [ 0, [  ] ] ]

      gap> m4_15.AutomorphismGroup;
      C3
130
      gap> m4_15.IsManifold;
      #I  SCIsManifold: processing vertex link 1/15
      #I  round 0: [ 13, 42, 58, 29 ]
      #I  round 1: [ 12, 38, 52, 26 ]
135 #I  round 2: [ 11, 34, 46, 23 ]
      #I  round 3: [ 10, 30, 40, 20 ]
      #I  round 4: [ 9, 26, 34, 17 ]
      #I  round 5: [ 8, 22, 28, 14 ]
      #I  round 6: [ 7, 18, 22, 11 ]
140 #I  round 7: [ 6, 14, 16, 8 ]
      #I  round 8: [ 5, 10, 10, 5 ]
      #I  SCReduceComplexEx: computed locally minimal complex after 9 rounds.
      #I  SCIsManifold: link is sphere.
      #I  SCIsManifold: processing vertex link 2/15
```

```
145 ...
    #I    SCIsManifold: processing vertex link 15/15
    #I    round 0: [ 13, 42, 58, 29 ]
    #I    round 1: [ 12, 38, 52, 26 ]
    #I    round 2: [ 11, 34, 46, 23 ]
150 #I    round 3: [ 10, 30, 40, 20 ]
    #I    round 4: [ 9, 26, 34, 17 ]
    #I    round 5: [ 8, 22, 28, 14 ]
    #I    round 6: [ 7, 18, 22, 11 ]
    #I    round 7: [ 6, 14, 16, 8 ]
155 #I    round 8: [ 5, 10, 10, 5 ]
    #I    SCReduceComplexEx: computed locally minimal complex after 9 rounds.
    #I    SCIsManifold: link is sphere.
    true

160 gap> m4_15.IsInKd(1);
    #I    SCIsInKd: checking link 1/15
    #I    SCIsKStackedSphere: try 1/50
    #I    round 0: [ 13, 42, 58, 29 ]
    #I    round 1: [ 12, 38, 52, 26 ]
165 #I    round 2: [ 11, 34, 46, 23 ]
    #I    round 3: [ 10, 30, 40, 20 ]
    #I    round 4: [ 9, 26, 34, 17 ]
    #I    round 5: [ 8, 22, 28, 14 ]
    #I    round 6: [ 7, 18, 22, 11 ]
170 #I    round 7: [ 6, 14, 16, 8 ]
    #I    round 8: [ 5, 10, 10, 5 ]
    #I    SCReduceComplexEx: computed locally minimal complex after 9 rounds.
    #I    SCIsInKd: checking link 2/15
    ...
175 #I    SCIsInKd: checking link 15/15
    #I    SCIsKStackedSphere: try 1/50
    #I    round 0: [ 13, 42, 58, 29 ]
    #I    round 1: [ 12, 38, 52, 26 ]
    #I    round 2: [ 11, 34, 46, 23 ]
180 #I    round 3: [ 10, 30, 40, 20 ]
    #I    round 4: [ 9, 26, 34, 17 ]
    #I    round 5: [ 8, 22, 28, 14 ]
    #I    round 6: [ 7, 18, 22, 11 ]
    #I    round 7: [ 6, 14, 16, 8 ]
185 #I    round 8: [ 5, 10, 10, 5 ]
    #I    SCReduceComplexEx: computed locally minimal complex after 9 rounds.
    #I    SCIsInKd: all links are 1-stacked.
    1
    gap> SCInfoLevel(2);
190 gap> m4_15.IsTight;
    #I    SCIsTight: complex is in class K(1) and 2-neighborly, thus tight.
    true
```

```
gap> PrintArray(SCMorseMultiplicityVector(m4_15,[1..15]));
     [ [   1,   0,   0,   0,   0 ],
195    [   0,   0,   0,   0,   0 ],
       [   0,   0,   0,   0,   0 ],
       [   0,   0,   0,   0,   0 ],
       [   0,   0,   0,   1,   0 ],
       [   0,   0,   0,   1,   0 ],
200    [   0,   1,   0,   0,   0 ],
       [   0,   0,   0,   0,   0 ],
       [   0,   1,   0,   0,   0 ],
       [   0,   0,   0,   1,   0 ],
       [   0,   1,   0,   0,   0 ],
205    [   0,   0,   0,   0,   0 ],
       [   0,   0,   0,   0,   0 ],
       [   0,   0,   0,   0,   0 ],
       [   0,   0,   0,   0,   1 ] ]
```

Appendix D

Enumeration algorithm for the 24-cell

The following **GAP** script is also available in digital form on the author's website [41] and upon request.

```
1  ################################################################################
   ################################################################################
   ####                      surface24cell.gap                      ####
   ################################################################################
5  ################################################################################
   #### Author: Felix Effenberger , 2008                            ####
   ####                                                             ####
   #### Description:                                                ####
   #### This program constructs all possible 2-dim. subcomplexes of ####
10 #### Skel_2(24-cell) fulfilling the pseudomanifold-property and inducing ####
   #### a Hamiltonian or split (singular vertex, two 4-cycles) path in the ####
   #### link of each vertex of the 24-cell , i.e. all Hamiltonian (pinch point) ####
   #### surfaces of Skel_2(24-cell).                                ####
   ####                                                             ####
15 #### This is accomplished by the following procedure:            ####
   ####                                                             ####
   #### The algorithm recursively tries all possibilities to kill 32 of the 96 ####
   #### triangles Skel_2(24-cell) respecting the restrictions metioned above ####
   #### yielding 2d complexes with Euler-Characteristic \Chi=-8.    ####
20 ####                                                             ####
   #### Remember that for each vertex v the link lk(v) is a cube and that a ####
   #### Hamiltonian surface yields hamiltonian paths in the links of all ####
   #### vertices of the 24-cell. In the cube there only exists one Hamil- ####
   #### tonian path modulo symmetries.                              ####
25 ####                                                             ####
   #### Since pp-surfaces are considered the program also deals with the ####
   #### case of two disjoint cycles of length 4 in lk(v). Note that this is ####
   #### the only valid splitting of the cube into disjoint paths in this case. ####
   ####                                                             ####
```

153

```
30 #### In any of the two cases, in each vertex figure (a cube), eight edges    ####
   #### (identified with triangles of Skel_2(24-cell)) belong to the complex    ####
   #### and four do not belong to the complex. These 4 triangles are marked as ####
   #### "killed", (not part of the surface) the other eight are marked as        ####
   #### "fixed" (part of the surface). Thus, a Hamiltonian surface can be ob-    ####
35 #### tained by constructing Hamiltonian or split paths in the vertex fi-      ####
   #### gures of all vertices and looking at the identified triangles of the     ####
   #### edges of all those paths.                                                ####
   ####                                                                          ####
   #### The algorithm works in a stepwise manner processing one vertex link      ####
40 #### after the other.                                                         ####
   ####                                                                          ####
   #### In the first step a path in the link of vertex 1 is fixed to be of       ####
   #### Hamiltonian or split type and then the first four triangles are          ####
   #### killed, the first eight fixed. As each edge of Skel_2(24-Cell) is con-  ####
45 #### tained in exactly three triangles of Skel_2(24-Cell) and for a           ####
   #### (pseudo)surface this number has to be two for each killed triangle one ####
   #### can now find two triangles that must be included in the (pseudo)         ####
   #### surface. These triangles now fix edges in the links of other vertices    ####
   #### (so called "associated" vertices), reducing the number of possibilities####
50 #### of Hamiltonian or split paths in the links of those vertices. For the    ####
   #### rest of the links all possibilities of Hamiltonian and split paths       ####
   #### in the links are tested, taking into account the growing number of       ####
   #### restrictions due to the fixed and killed triangles caused by the pre-    ####
   #### vious steps. If it is impossible to find a Hamiltonian path in lk(v)     ####
55 #### of a vertex v due to the killed and fixed edges configuration in lk(v) ####
   #### induced by the previous steps the construction will not result in a      ####
   #### surface and can be discarded.                                            ####
   ####                                                                          ####
   #### The algorithm makes use of this fact and systematically enumerates       ####
60 #### all possibilities to construct a Hamiltonian (pseudo)surface as           ####
   #### subcomplex of Skel_2(24-cell) using a backtracking algorithm.            ####
   ####                                                                          ####
   #### The program produces textual output to be able to see what the algo-     ####
   #### rihm is computing. The output is printed to the screen and also          ####
65 #### written to the file "surface24cell.log"                                  ####
   ####                                                                          ####
   ###   All found complexes are saved to output files of the form              ####
   #### "psurf24_X.dat", where X is a consecutive number starting at 1.          ####
   #### These files are all in GAP format and contain the list of               ####
70 #### simplices of the complex in the variable complex:=... and a list of      ####
   #### the links for all vertices 1-24 in the variable links:=... -- here       ####
   #### links[1]=lk(1), etc.                                                     ####
   ####                                                                          ####
   ################################################################################
75 ####                                                                          ####
   #### Tested with GAP Version 4.4.9                                            ####
   ####                                                                          ####
```

```
    ################################################################
    ################################################################
80
    LogTo("surface24cell.log");

    ################################################################
    ################################################################
85  ####                    GLOBAL VARIABLES                    ####
    ################################################################
    ################################################################


    ### trig: the 96 triangles of skel_2(24-cell)
90  trig :=
    [ [ 1, 2, 3 ], [ 1, 2, 4 ], [ 1, 2, 5 ], [ 1, 3, 6 ], [ 1, 3, 7 ],
      [ 1, 4, 6 ], [ 1, 4, 9 ], [ 1, 5, 7 ], [ 1, 5, 9 ], [ 1, 6, 11 ],
      [ 1, 7, 11 ], [ 1, 9, 11 ], [ 2, 3, 8 ], [ 2, 3, 10 ], [ 2, 4, 8 ],
      [ 2, 4, 12 ], [ 2, 5, 10 ], [ 2, 5, 12 ], [ 2, 8, 13 ], [ 2, 10, 13 ],
95    [ 2, 12, 13 ], [ 3, 6, 8 ], [ 3, 6, 14 ], [ 3, 7, 10 ], [ 3, 7, 14 ],
      [ 3, 8, 15 ], [ 3, 10, 15 ], [ 3, 14, 15 ], [ 4, 6, 8 ], [ 4, 6, 16 ],
      [ 4, 8, 17 ], [ 4, 9, 12 ], [ 4, 9, 16 ], [ 4, 12, 17 ], [ 4, 16, 17 ],
      [ 5, 7, 10 ], [ 5, 7, 18 ], [ 5, 9, 12 ], [ 5, 9, 18 ], [ 5, 10, 19 ],
      [ 5, 12, 19 ], [ 5, 18, 19 ], [ 6, 8, 20 ], [ 6, 11, 14 ], [ 6, 11, 16 ],
100   [ 6, 14, 20 ], [ 6, 16, 20 ], [ 7, 10, 21 ], [ 7, 11, 14 ], [ 7, 11, 18 ],
      [ 7, 14, 21 ], [ 7, 18, 21 ], [ 8, 13, 15 ], [ 8, 13, 17 ], [ 8, 15, 20 ],
      [ 8, 17, 20 ], [ 9, 11, 16 ], [ 9, 11, 18 ], [ 9, 12, 22 ], [ 9, 16, 22 ],
      [ 9, 18, 22 ], [ 10, 13, 15 ], [ 10, 13, 19 ], [ 10, 15, 21 ],
      [ 10, 19, 21 ], [ 11, 14, 23 ], [ 11, 16, 23 ], [ 11, 18, 23 ],
105   [ 12, 13, 17 ], [ 12, 13, 19 ], [ 12, 17, 22 ], [ 12, 19, 22 ],
      [ 13, 15, 24 ], [ 13, 17, 24 ], [ 13, 19, 24 ], [ 14, 15, 20 ],
      [ 14, 15, 21 ], [ 14, 20, 23 ], [ 14, 21, 23 ], [ 15, 20, 24 ],
      [ 15, 21, 24 ], [ 16, 17, 20 ], [ 16, 17, 22 ], [ 16, 20, 23 ],
      [ 16, 22, 23 ], [ 17, 20, 24 ], [ 17, 22, 24 ], [ 18, 19, 21 ],
110   [ 18, 19, 22 ], [ 18, 21, 23 ], [ 18, 22, 23 ], [ 19, 21, 24 ],
      [ 19, 22, 24 ], [ 20, 23, 24 ], [ 21, 23, 24 ], [ 22, 23, 24 ] ];


    ### edges: the 96 edges of of skel_2(24-cell)
115 edges:=[ [ 1, 2 ], [ 1, 3 ], [ 1, 4 ], [ 1, 5 ], [ 1, 6 ], [ 1, 7 ], [ 1, 9 ],
      [ 1, 11 ], [ 2, 3 ], [ 2, 4 ], [ 2, 5 ], [ 2, 8 ], [ 2, 10 ], [ 2, 12 ],
      [ 2, 13 ], [ 3, 6 ], [ 3, 7 ], [ 3, 8 ], [ 3, 10 ], [ 3, 14 ], [ 3, 15 ],
      [ 4, 6 ], [ 4, 8 ], [ 4, 9 ], [ 4, 12 ], [ 4, 16 ], [ 4, 17 ], [ 5, 7 ],
      [ 5, 9 ], [ 5, 10 ], [ 5, 12 ], [ 5, 18 ], [ 5, 19 ], [ 6, 8 ], [ 6, 11 ],
120   [ 6, 14 ], [ 6, 16 ], [ 6, 20 ], [ 7, 10 ], [ 7, 11 ], [ 7, 14 ],
      [ 7, 18 ], [ 7, 21 ], [ 8, 13 ], [ 8, 15 ], [ 8, 17 ], [ 8, 20 ],
      [ 9, 11 ], [ 9, 12 ], [ 9, 16 ], [ 9, 18 ], [ 9, 22 ], [ 10, 13 ],
      [ 10, 15 ], [ 10, 19 ], [ 10, 21 ], [ 11, 14 ], [ 11, 16 ], [ 11, 18 ],
      [ 11, 23 ], [ 12, 13 ], [ 12, 17 ], [ 12, 19 ], [ 12, 22 ], [ 13, 15 ],
125   [ 13, 17 ], [ 13, 19 ], [ 13, 24 ], [ 14, 15 ], [ 14, 20 ], [ 14, 21 ],
```

```
       [ 14, 23 ], [ 15, 20 ], [ 15, 21 ], [ 15, 24 ], [ 16, 17 ], [ 16, 20 ],
       [ 16, 22 ], [ 16, 23 ], [ 17, 20 ], [ 17, 22 ], [ 17, 24 ], [ 18, 19 ],
       [ 18, 21 ], [ 18, 22 ], [ 18, 23 ], [ 19, 21 ], [ 19, 22 ], [ 19, 24 ],
       [ 20, 23 ], [ 20, 24 ], [ 21, 23 ], [ 21, 24 ], [ 22, 23 ], [ 22, 24 ],
130    [ 23, 24 ]  ];


    ### global variables needed by the backtracking algorithm
    numkilled:=0;
135 mat:=[];
    killedrows:=[];
    numedget:=[];
    minrow:=1;
    backtrackstatusvec:=[];
140 surfcollection:=[];
    numsurfs:=0;
    startcallidx:=1;
    startcallidxdepth:=[];
    startrow:=[];
145 toplink:=[];
    linktrig:=[];
    linktrigidx:=[];


    ################################################################
150 ################################################################
    ####                      FUNCTIONS                        ####
    ################################################################
    ################################################################

155 ### computeLinks ##############################################
    # returns the link of every vertex in the given complex
    #
    computeLinks:=function(complex)
      local i,simplex,linkSimplex,link;
160
      link:=[];
      for i in [1..24] do
        link[i]:=[];

165      for simplex in complex do
           if i in simplex then
             linkSimplex:=ShallowCopy(simplex);
             RemoveSet(linkSimplex,i);
             AddSet(link[i],linkSimplex);
170        fi;
        od;
      od;
      return link;
```

```
    end;
175

    ### getTrianglesEdge ##########################################################
    # returns a list of triangles a given edge is contained in
    #
180 getTrianglesEdge:=function(edge)
      local t,list;
      list:=[];
      for t in trig do
        if(IsSubset(t,edge)) then
185       Add(list,t);
        fi;
      od;
      return list;
    end;
190


    ### getKilledLinkEdges ##########################################################
    # returns the list of edges that are killed in link lk(v)
    #
195 getKilledLinkEdges:=function(v)
      local e,t,idx,killededges;
      killededges:=[];
            for e in toplink[v] do
        t:=Union(e,[v]); #triangle that consists of edge in link+inner vertex
200     idx:=Position(trig,t);

        if(idx=fail) then #should never happen
          Print("error in getKilledLinkEges: triangle ",t," not found!\n");
          return [];
205     fi;

        if(killedrows[idx]=1) then
          Add(killededges,e); #killed edge
        fi;
210   od;
      return killededges;
    end;


215 ### getGraphCycle ##########################################################
    # determines, whether a given graph (as subset of the graph of a cube)
    # has a cycle or not
    #
    getGraphCycle:=function(curv,lastv,cyc,graph)
220   local poss,hascyc,e,nv;
```

157

```
     poss:=Filtered(graph,x->(curv in x and not lastv in x));

     hascyc:=0;
225  for e in poss do
       nv:=Difference(e,[curv])[1];
       if(cyc[nv]=1) then
         return 1; #found cycle
       fi;
230  od;

     for e in poss do
       nv:=Difference(e,[curv])[1];
       cyc[nv]:=1;
235    if(getGraphCycle(nv,curv,cyc,graph)=1) then
         return 1; #found cycle
       fi;
       cyc[nv]:=0;
     od;
240
     return 0; #no cycle
   end;


245 ### pathHasForbiddenVertices ##################################################
   # determines, whether a given path has "forbidden" vertices, i.e. vertices
   # with a degree of 3
   #
   pathHasForbiddenVertices:=function(path)
250  local deg,e;

     deg:=ListWithIdenticalEntries(24,0);
     for e in path do
       deg[e[1]]:=deg[e[1]]+1;
255    deg[e[2]]:=deg[e[2]]+1;
     od;

     if(3 in deg) then
       return 1;
260  else
       return 0;
     fi;
   end;

265
   ### getHamiltonBacktrack ######################################################
   # helper function for getHamiltonPaths - backtracking algorithm that computes
   # all possible hamilton paths in a cube containing the path p
   #
```

```
270  getHamiltonBacktrack:=function(v,curv,markedv,path,allp)
        local possnext,nv,e,ee,cyc,cur;

        #found hamiltonian path in the cube
        if(Length(path)=8) then
275       AddSet(allp,ShallowCopy(path));
          return;
        fi;

        #calculate possible next vertices (edges)
280     possnext:=[];
        for nv in [1..24] do
          e:=Set([curv,nv]);
          if(e in toplink[v] and not e in path) then

285         #cycle & branch detection
            cyc:=ListWithIdenticalEntries(24,0);
            cyc[curv]:=cyc[curv]+1;
            cyc[nv]:=cyc[nv]+1;
            for ee in path do
290           cyc[ee[1]]:=cyc[ee[1]]+1;
              cyc[ee[2]]:=cyc[ee[2]]+1;
            od;

            if(3 in cyc) then
295           #found cycle
              continue;
            fi;

            cyc:=ListWithIdenticalEntries(24,0);
300         cyc[curv]:=1;
            cyc[nv]:=1;
            if(getGraphCycle(curv,nv,cyc,Union(path,[Set([curv,nv])]))=1) then
              if(Length(path)=7) then
                AddSet(allp,Union(path,[e]));
305           fi;
              continue; #no cycle of length < 8 allowed
            fi;

            AddSet(possnext,nv);
310       fi;
        od;

        #recurse for next possibilities
        for nv in possnext do
315       markedv[nv]:=1;

          AddSet(path,Set([curv,nv]));
```

```
       getHamiltonBacktrack(v,nv,markedv,path,allp);
       RemoveSet(path,Set([curv,nv]));
320    markedv[nv]:=0;
     od;
   end;


325 ### getHamiltonPaths #####################################################
   # returns all Hamiltonian paths in the cube lk(v) that contain the given path
   # fixe in lk(v). returns [] if no such paths exist
   #
   getHamiltonPaths:=function(v,fixe)
330    local markedv,e,allp,p,cyc;

     #check for forbidden vertices of fixe
     if(pathHasForbiddenVertices(fixe)=1) then
       return []; #no hamilton path possible
335    fi;

     #check for cycles of fixe
     cyc:=ListWithIdenticalEntries(24,0);
     p:=ShallowCopy(fixe);
340    cyc[p[1][1]]:=1;
     cyc[p[1][2]]:=1;
     if(getGraphCycle(p[1][1],p[1][2],cyc,p)=1) then
       return []; #no hamilton path possible
     fi;
345
     #mark vertices of first edge of fixe
     markedv:=ListWithIdenticalEntries(24,0);
     markedv[fixe[1][1]]:=1;
     markedv[fixe[1][2]]:=1;
350
     #fix first working edge, then start backtrack
     allp:=[];
     p:=[fixe[1]];
     getHamiltonBacktrack(v,fixe[1][1],markedv,p,allp);
355
     #now return only return paths that have fixe as subset
     return Filtered(allp,x->IsSubset(x,Set(fixe)));
   end;


360
   ### cartesianEdges #######################################################
   # helper function, computes the cartesian product of a given set of edges.
   # returns the edges defined via the cartesian product of the vertex sets of the
   # parameter edges
365 #
```

```
    cartesianEdges:=function(edges)
      local all, alls, c;

      all:=[];
370   for c in Combinations(edges,2) do
        UniteSet(all, Cartesian(c));
      od;

      alls:=[];
375   for c in all do
        if(Length(Set(c))<2) then continue; fi;
        AddSet(alls, Set(c));
      od;

380   return Union(alls, edges);
    end;


    ### getSplitPaths ###############################################################
385 # returns all possible "split paths" in the link lk(v) of a vertex v containing
    # the set of edges fixe as subset.
    # here a "split path" is a set of two disjoint cycles of length 4 in the graph
    # of a cube
    #
390 getSplitPaths:=function(v,fixe)
      local i,e,f,vert,vert2,e1,e2,cand,overt,vertices,paths,edges,allcand,failed;

      #no path possible
      if(Length(fixe)>8 or pathHasForbiddenVertices(fixe)=1) then
395     return [];
      fi;

      #check whether edges intersect in one vertex
      e1:=[];
400   e2:=[];
      vert:=0;
      for e in fixe do
        for f in fixe do
          if(e=f) then continue; fi;
405       if(Intersection(e,f)<>[]) then
            vert:=Intersection(e,f)[1];
            e1:=e;
            e2:=f;
            break;
410       fi;
        od;
        if(vert<>0) then break; fi;
      od;
```

```
415   #4 cases
      if (vert <>0) then
        #first case - two edges intersect in one vertex
        #find 4th vertex
        cand:= Filtered (toplink [v] ,
420                    x->Intersection (x,e1)<>[] and x<>e1 and x<>e2);
        UniteSet (cand, Filtered (toplink [v] ,
                       x->Intersection (x,e2)<>[] and x<>e1 and x<>e2 ));

        vert2 := [];
425     for e in cand do
          for f in cand do
            if (f=e) then continue;  fi;
            if (Length (Intersection (e,f))=1) then
              UniteSet (vert2 , Intersection (e,f));
430         fi ;
          od;
        od;

        #4 vertices on one side
435     vertices:=Union(e1 ,e2 );
        UniteSet (vertices , Union ([vert] , vert2 ));

        #complementary vertices
        overt := [];
440     for e in toplink [v] do
          UniteSet (overt ,e);
        od;

        overt:= Difference (overt , vertices );
445
        #check whether there exist edges linking two sides -> forbidden
        failed :=0;
        for e in fixe do
          if ((e[1] in vertices and e[2] in overt) or
450           (e[2] in vertices and e[1] in overt)) then
            failed :=1;
          fi ;
        od;

455     if (failed =1) then
          return  [];
        else
          #two 4-cycles
          cand:= Filtered (toplink [v] ,x->IsSubset (vertices ,x ));
460       UniteSet (cand , Filtered (toplink [v] ,x->IsSubset (overt ,x)));
          return  [cand];
```

```
         fi ;
     else
         #second to forth case -- edges disjoint
465      paths:=Filtered(toplink[v],x->Length(Intersection(x,fixe[1]))=1);

         edges:=[];
         for e in cartesianEdges(paths) do
           if(e in toplink[v] and e<>fixe[1] and
470            Intersection(e,fixe[1])=[]) then
             AddSet(edges,e);
           fi ;
         od ;

475      #two possible orientations

         allcand:=[];
         for i in [1..2] do
           #4 vertices on one side
480          vertices:=Union(fixe[1],edges[i]);

           #complementary vertices
           overt:=[];
           for e in toplink[v] do
485            UniteSet(overt,e);
           od ;

           overt:=Difference(overt,vertices);

490        #check whether there exist edges linking two sides -> forbidden
           failed:=0;
           for e in fixe do
             if((e[1] in vertices and e[2] in overt) or
                (e[2] in vertices and e[1] in overt)) then
495              failed:=1;
             fi ;
           od ;

           if(failed=1) then
500          continue;
           else
             #two 4-cycles
             cand:=Filtered(toplink[v],x->IsSubset(vertices,x));
             UniteSet(cand,Filtered(toplink[v],x->IsSubset(overt,x)));
505          AddSet(allcand,cand);
           fi ;
         od ;

         return allcand;
```

```
510    fi ;
    end ;



    ### getSplitPaths ###############################################################
515 # called by the global backtracking algorithm when a (pseudo-)manifold is
    # found. here it is saved to the global list surfcollection.
    # The number of found complexes in total is saved to numsurf
    #
    savesurf:=function ()
520    local i , surf , file , links , l ,hom;
      surf := [];
      for i in [1.. Length( trig )] do
        if( killedrows [i]=1) then continue; fi ;
        AddSet( surf , trig [ i ]);
525    od;

      if( surf in surfcollection ) then
        return ; #no doubles
      fi ;
530
      numsurfs:=numsurfs+1;
      Add( surfcollection , ShallowCopy ( surf ));

      file := Concatenation ([" psurf24_ ", String ( numsurfs ),". dat" ]);
535    PrintTo ( file ," complex:=", surf ," ;; \ n\n" );
      links :=computeLinks ( surf );
      AppendTo ( file ," links:=", links ," ;; \ n\n" );

      hom := [];
540    for l in [1.. Length ( links )] do
        #disable the following comment to enable
        #homology computation for the links

        #AppendTo ( file ," #", l ," - ", SimplicialHomology ( links [ l ]),"\n" );
545    od;

    end ;



550 ### isValidPath ###############################################################
    # helper function for killTrianglesLink , determines whether a given path p in
    # the link lk(v) of a vertex v is valid with respect to the already killed and
    # fixed triangles -- i.e. it must not contain killed triangles and must contain
    # all fixed triangles related to that path.
555 #
    isValidPath:=function (v,p)
      local e ,t , pt , ptc ;
```

```
       pt := [];
560    for e in p do
         t := Position(trig, Union(e, [v]));
         AddSet(pt, t);

         if(killedrows[t]=1) then
565        return false;
         fi;
       od;

       ptc := Difference(linktrigidx[v], pt);
570
       for t in ptc do
         if(killedrows[t]=2) then
           return false;
         fi;
575    od;

       return true;
     end;

580
     ### killTrianglesLink ##########################################################
     # helper function for getNextLink. enumerates all possibilities of split paths
     # and Hamiltonian paths in the link lk(v) of a vertex v respecting the set of
     # fixed edges fixededges that have to be part of the paths.
585  # returns the a list of triangles to be killed and one of triangles to be fixed
     # for the idx-th path of all those paths or [] if either no such paths exists
     # or the the number of such paths is <idx
     #
     # note: modulo symmetries of the cube there is only one possibility for a
590  # Hamiltonian path in the cube:
     #                           .
     #       h . _____ g
     #       |       /
     #       |      /
595  # e . _____/ f
     #     |  |
     #     |d | _____ c
     #     |       /
     #     |      /
600  # a | _____/ b
     #
     # similarly, there is only one possibility for a split path:
     #
     #       h. _____ g
605  #       /       /
```

165

```
    #     /          /
    #   e/ _____ /f
    #
    #     d _____ c
610 #     /          /
    #    /          /
    #   a/ _____ /b
    #
    #
615 killTrianglesLink:=function(v,fixededges,idx)
      local hpaths,spaths,allpaths,e,tokill,tofix;

      #get all hamiltonian paths and all split paths respecting fixededges
      hpaths:=getHamiltonPaths(v,fixededges);
620   spaths:=getSplitPaths(v,fixededges);

      allpaths:=Union(hpaths,spaths);

      #extract valid paths
625   allpaths:=Filtered(allpaths,x->isValidPath(v,x));

      if(idx>Length(allpaths)) then
        return [];
      fi;
630
      #construct sets of killed & fixed triangles
      tokill:=[];
      for e in Difference(toplink[v],allpaths[idx]) do
        AddSet(tokill,Union(e,[v]));
635   od;

      tofix:=[];
      for e in allpaths[idx] do
        AddSet(tofix,Union(e,[v]));
640   od;

      if(Length(tokill)<>4) then
        #should never happen
        Print("killTrianglesLink: error! length tokill=",Length(tokill),"!\n");
645   fi;

      #return list of triangles to be killed
      return [tokill,tofix];
    end;
650

    ### getPuzzleVertex ############################################################
    # helper function for constructComplexBacktrack.
```

```
      #
655   # let  lk(v)  be  an  already  processed  vertex  link  with  4  killed ,  8  fixed  edges.
      # an  associated  vertex  w_i  is  a  vertex ,  whose  link  contains  at  least  2  edges
      # that  are  already  killed  in  lk(v)  and  therefore  have  to  be  contained  in
      # lk(w_i).
      #
660   # the  function  returns  the  idx-th  associated  vertex  w_i  of  v  along  with  edges
      # that  are  killed  in  lk(v)  and  must  be  contained  in  lk(w_i)
      #
      # note  that  a  triangulated  (pseudo)surface  fulfills  the  pm-property ,  i.e.  every
      # edge  e  is  contained  in  exactly  two  triangles.  If  one  triangle  in  trig
665   # containing  an  edge  e  is  already  killed  exactly  two  triangles  that  contain  e
      # are  left  and  both  have  to  be  in  the  surface  (for  a  Hamiltonian  surface
      # contains  every  edge  of  the  96  edges  of  the  24cell).
      #
      getPuzzleVertex:=function(v,killededges ,idx)
670     local  e,e1,e2,t,tt,linkedv,i,j,lidx;
        lidx:=idx;

          #for  all  pairs  of  killed  edges
        for  e1  in  [1..Length(killededges)-1]  do
675       for  e2  in  [e1+1..Length(killededges)]  do
            #get  all  triangles  that  include  e1  or  e2
            t:=[];
            t[1]:=getTrianglesEdge(killededges[e1]);
            t[2]:=getTrianglesEdge(killededges[e2]);
680

            #extract  linked  vertices
            e:=[e1,e2];
            linkedv:=[];
            for  i  in  [1..2]  do
685           linkedv[i]:=[];
              #for  all  triangles  that  include  e_i  (i = 1,2)
                    for  tt  in  t[i]  do
                if(v  in  tt)  then
                   continue; #skip  triangles  in  current  link
690             fi ;
              UniteSet(linkedv[i],Difference(tt,killededges[e[i]]));
              od;
            od;

695         #look  for  vertex  that  is  linked  at  two  edges
            for  i  in  [1..Length(linkedv[1])]  do
              j:=Position(linkedv[2],linkedv[1][i]);
              if(j<>fail)  then
                lidx:=lidx-1;
                #return  idx-th  associated  vertex
700             if(lidx=0)  then
```

```
                 return
                 [linkedv[1][i],[killededges[e1],killededges[e2]]];
              fi;
705        fi;
        od;
      od;
    od;
    return []; #no more associated vertices
710 end;


    ### getKilledLinkEdges ########################################################
    # helper function for getNextLink. returns the list of edges that
715 # are killed in link lk(v) of vertex v
    #
    getKilledLinkEdges:=function(v)
      local e,t,idx,killededges;
      killededges:=[];
720        for e in toplink[v] do
        t:=Union(e,[v]); #triangle that consists of edge in link + inner vertex
        idx:=Position(trig,t);

        if(idx=fail) then
725        #should not happen
          Print("error in getKilledLinkEges: error! triangle ",
                t," not found!\n");
          return [];
        fi;
730
        if(killedrows[idx]=1) then
          #killed edge
          Add(killededges,e);
        fi;
735   od;
      return killededges;
    end;



740 ### getNextLink ###############################################################
    # helper function for constructComplexBacktrack. computes all possiblilities
    # of killing edges in the links of "associated" vertices of the vertex v.
    # see getPuzzleVertex for associated vertex.
    # the idx-th possibility is chosen and returned.
745 #
    getNextLink:=function(v,idx)
      local killedge,lidx,j,i,poss,poss2,cur,cur2,t,tidx,numk,countk;

      #get killed edges in lk(v) via global killedrows
```

```
750    killedge:=getKilledLinkEdges(v);

       #since v was already processed , exactly four edges of lk(v) are be
       #marked as killed edges
       if(Length(killedge)<>4) then
755      #should never happen
         Print("getNextPossibility: error! vertex ",v," not active!\n");
         return [];
       fi;


760    poss:=[];
       lidx:=1;
         #returns lidx-th "associated" vertex (cur[lidx][1]) (having at least
       #2 killed edges in its link
         #and two killed edges in vertex link lk(cur[lidx][1]) (cur[lidx][2])
765    cur:=getPuzzleVertex(v,killedge , lidx);
       while(cur<>[]) do
         AddSet(poss ,cur);
         lidx:=lidx+1;
         cur:=getPuzzleVertex(v,killedge , lidx);
770    od;


       #Size(poss) = number of associated vertices

       if(Length(poss)=0) then
775      #algorithm consistency check
         countk:=ListWithIdenticalEntries(24 ,0);
         for i in [1..24] do
           for j in linktrigidx[i] do
             if(killedrows[j]=1) then
780              countk[i]:=countk[i]+1;
             fi;
           od;
         od;

785      if(1 in countk) then
           #should never happen
           Print("getNextLink: error! no linked vertices .",
                 "but vertices which could be used.\n");
           Print("countk: ",countk,"\n");
790      fi;
       fi;

       #save all possibilities of paths (each associated
       poss2:=[];
795    for cur in poss do
         lidx:=1;
         cur2:=killTrianglesLink(cur[1],cur[2],lidx);
```

169

```
        while(cur2<>[]) do
          numk:=0;
800             #for all killed triangles
          for t in cur2[1] do
                  #check if triangle exists
            tidx:=Position(trig,t);
            if(tidx=fail) then
805           #should never happen
              Print("getNextLink: error! triangle ",t," not found!\n\n");
            fi;
                  #check if triangled has already been killed
            if(killedrows[tidx]=0) then numk:=numk+1; fi;
810       od;

          #if there where non-killed triangles, add them to poss2
          if(numk>0) then
            AddSet(poss2,[cur[1],cur2[1],cur2[2]]);
815       fi;
          lidx:=lidx+1;
          cur2:=killTrianglesLink(cur[1],cur[2],lidx);
        od;
      od;
820
    #Size(poss2) = number of possilities

    #idx too big
    if(idx>Length(poss2)) then
825     return [];
      fi;

      #4 triangles and a vertex
      return poss2[idx];
830 end;


### constructComplexBacktrack ##############################################
    # main backtracking algorithm - constructs all possible subcomplexes of the
835 # set of triangles of the 24-cell fulfilling the pseudomanifold property and
    # having in the link of each vertex either a hamiltonian path or a split path
    #
    # lvl = backtrack-level
    # curv = current vertex
840 # idx = idx-th possibility to choose
    #
    constructComplexBacktrack:=function(lvl,curv,idx)
      local i,callidx,pseudo,failed,t,tidx,waskilled,wasfixed,nextL;

845   Print("backtrack: lvl=",lvl," idx=",idx," curv=",curv,"\n");
```

170

```
        #get first candidates for triangles to be killed
     #nextL[1]= next vertex
     #nextL[2]= triangles to kill
850  #nextL[3]= triangles to fix
     nextL:=getNextLink(curv,idx);

     Print("backtrack: curv=",curv," nextl=",nextL,"\n");

855  if(nextL=[]) then
        #no more possibilities, step back
        Print("backtrack: no more possibilities, terminating branch.\n\n");
        return -1;
     fi;
860
     #kill triangles
     waskilled:=[];
     wasfixed:=[];
     for t in nextL[2] do
865     tidx:=Position(trig,t);
        if(tidx=fail) then
          #should never happen
          Print("backtrack: error! couldn't find triangle ",
               t," in triangulation!\n");
870        return -1;
        fi;

        if(killedrows[tidx]=0) then
          Add(waskilled,tidx);
875       killedrows[tidx]:=1;
          for i in [1..Length(numedget)] do
            numedget[i]:=numedget[i]-mat[tidx][i];
          od;
        fi;
880  od;

     for t in nextL[3] do
        tidx:=Position(trig,t);
        if(tidx=fail) then
885       #should never happen
          Print("backtrack: error! couldn't find triangle ",
               t," in triangulation!\n");
          return -1;
        fi;
890     killedrows[tidx]:=2;
        AddSet(wasfixed,tidx);
     od;
```

171

```
     Print("backtrack: killed ",Length(waskilled),
895        " triangles in lk(",nextL[1],")\n");
     numkilled:=numkilled+Length(waskilled);

     pseudo:=1;
     failed:=0;
900  for i in [1..Length(numedget)] do
       if(numedget[i]<2) then failed:=1; break; fi;
       if(numedget[i]<>2) then pseudo:=0; fi;
     od;

905  if(failed=0) then
       Print("backtrack: valid triangulation, recursing.\n\n");
       if(pseudo=0) then
         if(numkilled<32) then
           callidx:=1;
910        while(constructComplexBacktrack(lvl+1,nextL[1], callidx)>=0) do
             callidx:=callidx+1;
           od;
         fi;
       else
915      #examine case
         Print("\nbacktrack: found pseudomanifold for numkilled=",
               numkilled,"\n\n");
         if(numkilled=32) then savesurf(); fi;
       fi;
920  else
       Print("backtrack: lvl=",lvl," idx=",idx,
             " - invalid triangulation, stepping back.\n");
     fi;

925  #unkill triangles
     numkilled:=numkilled-Length(waskilled);
     for tidx in waskilled do
       killedrows[tidx]:=0;
       for i in [1..Length(numedget)] do
930      numedget[i]:=numedget[i]+mat[tidx][i];
       od;
     od;

     #unfix triangles
935  for tidx in wasfixed do
       killedrows[tidx]:=0;
     od;

     Print("backtrack: lvl=",lvl," idx=",idx,
940        " - branch done, rebranch one up.\n\n");
```

172

```
    return 0;
  end;
```

```
  ###############################################################################
  ###############################################################################
  ####                      MAIN PROGRAM                      ####
  ###############################################################################
```

```
  ###############################################################################

  #compute the links of all triangles
  toplink:=computeLinks(trig);
  for i in [1..Length(toplink)] do
    linktrig[i]:=[];
    linktrigidx[i]:=[];
    for e in toplink[i] do
      t:=Union(Set(e),[i]);
      AddSet(linktrig[i],t);
      AddSet(linktrigidx[i],Position(trig,t));
    od;
  od;


  #two possible starting configurations in the link of vertex 1
  #note that the labeling in the first link can be freely chosen
  startconfigs:=[
  [ #hamilton path [2,3,6,11,7,5,9,4]
  #killed triangles (not in complex)
  [ [ 1, 2, 5 ], [ 1, 3, 7 ], [ 1, 4, 6 ], [ 1, 9, 11 ] ],
  #fixed triangles (in complex)
  [ [ 1, 2, 3 ], [ 1, 2, 4 ], [ 1, 3, 6 ], [ 1, 4, 9 ], [ 1, 5, 7 ], [ 1, 5, 9 ],
    [ 1, 6, 11 ], [ 1, 7, 11 ] ]
  ],
  [ #split path, two cycles [2,3,6,4] and [5,7,11,9]
  #killed triangles (not in complex)
  [ [ 1, 2, 5 ], [ 1, 3, 7 ], [ 1, 4, 9 ], [ 1, 6, 11 ] ],
  #fixed triangles (in complex)
  [ [ 1, 2, 3 ], [ 1, 2, 4 ], [ 1, 3, 6 ], [ 1, 4, 6 ], [ 1, 5, 7 ], [ 1, 5, 9 ],
    [ 1, 7, 11 ], [ 1, 9, 11] ]
  ]
  ];

  for startcase in [1..Length(startconfigs)] do
    startcasearr:=startconfigs[startcase];
    Print("--> stating calculation for case ",
          startcase," ( ", startcasearr," ) in lk(1).\n\n");


    #calculate incidence matrix
```

955

960

965

970

975

980

985

```
990    #matrix[i][j] = 1 iff edges[j] edge of trig[i]
       #else matrix[i][j] = 0

       #numedget[i]=number of triangles that contain edge[i]
       numedget:=ListWithIdenticalEntries(Length(edges),0);
995
       #setup matrix
       mat:=[];
       for tidx in [1..Length(trig)] do
         mat[tidx]:=[];
1000     for eidx in [1..Length(edges)] do
           if(IsSubset(trig[tidx],edges[eidx])) then
             mat[tidx][eidx]:=1;
             numedget[eidx]:=numedget[eidx]+1;
           else
1005         mat[tidx][eidx]:=0;
           fi;
         od;
       od;

1010   #number of killed triangles
       numkilled:=0;

       #killedrows[i]=1 when triangle i was killed, 0 otherwise
       killedrows:=ListWithIdenticalEntries(Length(trig),0);
1015
       #sets of killed (not part of complex) and fixed (part of complex) triangles
       killed:=startcasearr[1];
       fixed:=startcasearr[2];

1020   #update killedrows and numedget for killed triangles
       for t in killed do
         tidx:=Position(trig,t);
         if(killedrows[tidx]=0) then
           killedrows[tidx]:=1;
1025       for i in [1..Length(numedget)] do
             numedget[i]:=numedget[i]-mat[tidx][i];
           od;
         fi;
       od;
1030   numkilled:=4;

       #update killedrows for fixed triangles
       for t in fixed do
         killedrows[Position(trig,t)]:=2;
1035   od;

       #start backtracking algorithm with current starting configuration in lk(1)
```

```
     scurpos:=1;
     ret:=0;
1040 while(ret>=0) do #ret>=0 means at least one more branch left
        Print("-----> starting for vertex 1, idx ",scurpos,".\n\n");
        ret:=constructComplexBacktrack(1,1,scurpos);
        scurpos:=scurpos+1;
        Print("----> stopping for vertex 1, idx ",scurpos,".\n\n");
1045 od;

     Print("----> no more possibilities for case ",startcase,".\n\n");
   od;

1050 Print("--> no more possibilities, all done. found ",
          numsurfs," pseudo surfaces.\n");

   LogTo();

1055 ###############################################################################
     ###############################################################################
     ###                              END                                        ###
     ###############################################################################
     ###############################################################################
```

Appendix E

GAP program constructing a conjectured series of $S^k \times S^k$

The following GAP script is also available in digital form on the author's website [41] and upon request.

```
1  ###############################################################################
   ###############################################################################
   ####                         seriesksks.gap                              ####
   ###############################################################################
5  ###############################################################################
   #### Author: Felix Effenberger, 2009                                     ####
   ####                                                                     ####
   #### Description:                                                        ####
   #### ------------                                                        ####
10 #### Constructs a series of triangulations M^{2(k-1)} which are conjectured ####
   #### to be (k-1)-Hamiltonian in the 2k-cross polytope and of the topological####
   #### type M^2(k-1)~=S^{k-1} x S^{k-1}.                                   ####
   ####                                                                     ####
   #### See Chapter 5 of F. Effenberger, Hamiltonian submanifolds of regular ####
15 #### polytopes, 2010, PhD thesis, University of Stuttgart for further de- ####
   #### tails.                                                              ####
   ####                                                                     ####
   #### This script needs the GAP package 'simpcomp' to run, which can be ob- ####
   #### tained at                                                           ####
20 ####                                                                     ####
   ####    http://www.gap-system.org/Packages/simpcomp.html                ####
   ####    or                                                              ####
   ####    http://www.igt.uni-stuttgart.de/LstDiffgeo/simpcomp/            ####
   ####                                                                     ####
25 #### Usage:                                                              ####
```

```
#### ------                                                            ####
#### Adjust the parameter 'kmax' in the global variables section below and  ####
#### execute the script. The output will be written to the screen and the  ####
#### log file 'seriessksk.log'. If the flag 'writefiles' in the global va-  ####
30 #### riables section is set to true, the complexes are written to the files ####
#### 'SKSK_{k}.sc', where {k} is the index in the series.                ####
####                                                                    ####
################################################################################
####                                                                    ####
35 #### Tested with GAP Version 4.4.9, simpcomp Version 1.1.21            ####
####                                                                    ####
################################################################################
################################################################################


40 LogTo("seriessksk.log");


################################################################################
################################################################################
####                        GLOBAL VARIABLES                            ####
45 ################################################################################
################################################################################


kmax:=5;              #maximal index to which the series should be constructed
writefiles:=false;  #flag to set whether output files with the triangulations
50                    #in simpcomp format (SKSK_{k}.sc) should be written


################################################################################
################################################################################
####                         MAIN PROGRAM                               ####
55 ################################################################################
################################################################################


#start with nncs 8 vertex triangulation of the torus
complex:=SCFromDifferenceCycles([[1,1,6],[3,3,2]]);
60
for k in [3..kmax] do

   Print("## Constructing case k=",k-1," -> k=",k," ##\n");

65 #number of vertices, labeled in Z/4kZ
   n:=4*k;

   #autormorphism group
   cyc:=PermList(Concatenation([2..n],[1]));          #cyclic generator
70 mula:=PermList((((0..n-1]*-1) mod n)+1);           #mult. *(-1)
   mulb:=PermList((((0..n-1]*(2*k-1)) mod n)+1);      #mult. *(2k-1)
   G:=Group(cyc,mula,mulb);
```

```
     #print automorphism group information
75   Print("Automorphism group G:\n|G|=",Size(G),
         "\nG~=",StructureDescription(G),"\n");
     Print("Generators:\n",cyc,"\n",mula,"\n",mulb,"\n\n");

     #map link from Z/4kZ to Z/4(k+1)Z
80   mapfacets:=[];
     for f in complex.Link(1).Facets do
       mf:=ShallowCopy(f)-1;
       for i in [1..Length(mf)] do
         if(mf[i]<2*(k-1)) then
85         # v mod 4*(k-1) -> v+1 mod 4*k
           mf[i]:=mf[i]+1;
         elif(mf[i]>2*(k-1)) then
           # -v mod 4*(k-1) -> -v-1 mod 4*k
           mf[i]:=mf[i]+3;
90       fi;
       od;

       #glue in simplex [-1,0,1]
       Add(mapfacets,Union(mf+1,[1,2,4*k]));
95   od;

     #generate complex from group operation
     complex:=SCFromFacets(Union(Orbits(G,Set(mapfacets),OnSets)));
     SCFaceLattice(complex); #speed up f-vector calculation
100  SCPropertySet(complex,"AutomorphismGroup",G);

     #print complex information
     Print("Complex:\nd=",complex.Dim,
         "\nF=",complex.F,
105        "\nChi=",complex.Chi,
         "\nHomology H_*=",complex.Homology,
         "\nGenerators:\n",complex.Generators,
         "\nGenerating difference cycles:\n",
         Set(List(complex.Generators,x->SCDifferenceCycleCompress(x[1],n))),
110        "\nAll difference cycles:\n",
         Set(List(complex.Facets,x->SCDifferenceCycleCompress(x,n))),
         "\n\n");

       #write files to disc
115    if(writefiles=true) then
         SCSave(complex,Concatenation("SKSK_",String(k),".sc"));
       fi;
   od;

120 LogTo();
```

APPENDIX E. GAP PROGRAM CONSTRUCTING A CONJECTURED SERIES OF $S^k \times S^k$

```
##############################################################################
##############################################################################
###                              END                                      ###
125 ##############################################################################
##############################################################################
```

Bibliography

[1] Ron M. Adin. *Combinatorial Structure of Simplicial Complexes with Symmetry*. PhD thesis, The Hebrew University, Jerusalem, 1991.

[2] Aleksandr D. Alexandrov. On a class of closed surfaces. *Recueil Math. (Moscow)*, 4:69–72, 1938.

[3] Amos Altshuler. Manifolds in stacked 4-polytopes. *J. Combinatorial Theory Ser. A*, 10:198–239, 1971.

[4] Amos Altshuler and Leon Steinberg. Neighborly 4-polytopes with 9 vertices. *J. Combinatorial Theory Ser. A*, 15:270–287, 1973.

[5] Amos Altshuler and Leon Steinberg. Neighborly combinatorial 3-manifolds with 9 vertices. *Discrete Math.*, 8:113–137, 1974.

[6] K. Appel and W. Haken. Every planar map is four colorable. I. Discharging. *Illinois J. Math.*, 21(3):429–490, 1977.

[7] K. Appel, W. Haken, and J. Koch. Every planar map is four colorable. II. Reducibility. *Illinois J. Math.*, 21(3):491–567, 1977.

[8] Bhaskar Bagchi and Basudeb Datta. Lower bound theorem for normal pseudomanifolds. *Expo. Math.*, 26(4):327–351, 2008.

[9] Bhaskar Bagchi and Basudeb Datta. Minimal triangulations of sphere bundles over the circle. *J. Combin. Theory Ser. A*, 115(5):737–752, 2008.

[10] Bhaskar Bagchi and Basudeb Datta. Uniqueness of Walkup's 9-vertex 3-dimensional Klein bottle. *Discrete Math.*, 308(22):5087–5095, 2008.

[11] Bhaskar Bagchi and Basudeb Datta. On Walkup's class $\mathcal{K}(d)$ and a minimal triangulation of a 4-manifold. arXiv:0804.2153v2 [math.GT], Preprint, 9 pages, 2010.

[12] Thomas F. Banchoff. Tightly embedded 2-dimensional polyhedral manifolds. *Amer. J. Math.*, 87:462–472, 1965.

[13] Thomas F. Banchoff. Tight polyhedral Klein bottles, projective planes, and Möbius bands. *Math. Ann.*, 207:233–243, 1974.

[14] Thomas F. Banchoff and Wolfgang Kühnel. Tight submanifolds, smooth and polyhedral. In *Tight and taut submanifolds (Berkeley, CA, 1994)*, volume 32 of *Math. Sci. Res. Inst. Publ.*, pages 51–118. Cambridge Univ. Press, Cambridge, 1997.

[15] Thomas F. Banchoff and Wolfgang Kühnel. Tight polyhedral models of isoparametric families, and PL-taut submanifolds. *Adv. Geom.*, 7(4):613–629, 2007.

[16] David Barnette. A proof of the lower bound conjecture for convex polytopes. *Pacific J. Math.*, 46:349–354, 1973.

[17] David Barnette. Graph theorems for manifolds. *Israel J. Math.*, 16:62–72, 1973.

[18] David W. Barnette. The minimum number of vertices of a simple polytope. *Israel J. Math.*, 10:121–125, 1971.

[19] Lowell W. Beineke and Frank Harary. The genus of the n-cube. *Canad. J. Math.*, 17:494–496, 1965.

[20] Thomas Beth, Dieter Jungnickel, and Hanfried Lenz. *Design theory. Vol. I*, volume 69 of *Encyclopedia of Mathematics and its Applications*. Cambridge University Press, Cambridge, second edition, 1999.

[21] Louis J. Billera and Carl W. Lee. A proof of the sufficiency of McMullen's conditions for f-vectors of simplicial convex polytopes. *J. Combin. Theory Ser. A*, 31(3):237–255, 1981.

[22] Anders Björner and Frank H. Lutz. Simplicial manifolds, bistellar flips and a 16-vertex triangulation of the Poincaré homology 3-sphere. *Experiment. Math.*, 9(2):275–289, 2000.

[23] G. Blind and R. Blind. Shellings and the lower bound theorem. *Discrete Comput. Geom.*, 21(4):519–526, 1999.

[24] Jürgen Bokowski and Anselm Eggert. Toutes les réalisations du tore de Möbius avec sept sommets. *Structural Topology*, (17):59–78, 1991.

[25] Raoul Bott. Morse theory and the Yang-Mills equations. In *Differential geometrical methods in mathematical physics (Proc. Conf., Aix-en-Provence/Salamanca, 1979)*, volume 836 of *Lecture Notes in Math.*, pages 269–275. Springer, Berlin, 1980.

[26] Raoul Bott. Lectures on Morse theory, old and new. *Bull. Amer. Math. Soc. (N.S.)*, 7(2):331–358, 1982.

[27] Ulrich Brehm and Wolfgang Kühnel. Combinatorial manifolds with few vertices. *Topology*, 26(4):465–473, 1987.

[28] Ulrich Brehm and Wolfgang Kühnel. 15-vertex triangulations of an 8-manifold. *Math. Ann.*, 294(1):167–193, 1992.

[29] Mario Casella and Wolfgang Kühnel. A triangulated $K3$ surface with the minimum number of vertices. *Topology*, 40(4):753–772, 2001.

[30] Thomas E. Cecil and Patrick J. Ryan. Tight and taut immersions into hyperbolic space. *J. London Math. Soc. (2)*, 19(3):561–572, 1979.

[31] Gary Chartrand and S. F. Kapoor. The cube of every connected graph is 1-hamiltonian. *J. Res. Nat. Bur. Standards Sect. B*, 73B:47–48, 1969.

[32] Shiing-Shen Chern and Richard K. Lashof. On the total curvature of immersed manifolds. *Amer. J. Math.*, 79:306–318, 1957.

[33] Jacob Chestnut, Jenya Sapir, and Ed Swartz. Enumerative properties of triangulations of spherical bundles over S^1. *European J. Combin.*, 29(3):662–671, 2008.

[34] Harold S. M. Coxeter. *Regular polytopes*. Dover Publications Inc., New York, third edition, 1973.

[35] Ákos Császár. A polyhedron without diagonals. *Acta Univ. Szeged. Sect. Sci. Math.*, 13:140–142, 1949.

[36] Basudeb Datta. Minimal triangulations of manifolds. *J. Indian Inst. Sci.*, 87(4):429–449, 2007.

[37] M. Dehn. Die Eulersche Formel im Zusammenhang mit dem Inhalt in der Nicht-Euklidischen Geometrie. *Math. Ann.*, 61(4):561–586, 1906.

[38] J.-G. Dumas, F. Heckenbach, B. D. Saunders, and V. Welker. Simplicial Homology, a GAP package, Version 1.4.3. http://www.cis.udel.edu/~dumas/Homology/, 2009.

[39] Herbert Edelsbrunner and John Harer. Persistent homology—a survey. In *Surveys on discrete and computational geometry*, volume 453 of *Contemp. Math.*, pages 257–282. Amer. Math. Soc., Providence, RI, 2008.

[40] Felix Effenberger. Hamiltonian subcomplexes of the 24-cell. Preprint, 3 pages with electronic geometry model, 2008. Submitted to *EG-Models*.

[41] Felix Effenberger. Personal website at the University of Stuttgart. http://www.igt.uni-stuttgart.de/LstDiffgeo/Effenberger/, 2010.

[42] Felix Effenberger. Stacked polytopes and tight triangulations of manifolds. arXiv:0911.5037v3 [math.GT], Preprint, 28 pages, 2010. Submitted for publication.

[43] Felix Effenberger and Wolfgang Kühnel. Hamiltonian submanifolds of regular polytopes. *Discrete Comput. Geom.*, 43(2):242–262, March 2010. Preprint available: arXiv:0709.3998v2 [math.CO].

[44] Felix Effenberger and Jonathan Spreer. simpcomp — A GAP package, Version 1.4.0. http://www.igt.uni-stuttgart.de/LstDiffgeo/simpcomp, 2010. Submitted to the *GAP Group*.

[45] Felix Effenberger and Jonathan Spreer. simpcomp — a GAP toolbox for simplicial complexes, Preprint, 4 pages. `arXiv:1004.1367v2 [math.CO]`, 2010. To appear in *ACM Commun. Comput. Algebra.*

[46] Richard Ehrenborg and Masahiro Hachimori. Non-constructible complexes and the bridge index. *European J. Combin.*, 22(4):475–489, 2001.

[47] Euclid. *Euclid's Elements*. Green Lion Press, Santa Fe, NM, 2002. All thirteen books complete in one volume, The Thomas L. Heath translation, Edited by Dana Densmore.

[48] Günter Ewald. Hamiltonian circuits in simplicial complexes. *Geometriae Dedicata*, 2:115–125, 1973.

[49] Michael Freedman. The topology of four-dimensional manifolds. *Journal of Differential Geometry*, 17:357–453, 1982.

[50] Michael Freedman and Robion Kirby. A geometric proof of Rohlin's theorem. In *Proc. Symp. Pure Math.*, volume 2, pages 85–97, 1978.

[51] GAP – Groups, Algorithms, and Programming, Version 4.4.12. `http://www.gap-system.org`, 2008.

[52] Ewgenij Gawrilow and Michael Joswig. polymake: a framework for analyzing convex polytopes. In *Polytopes—combinatorics and computation (Oberwolfach, 1997)*, volume 29 of *DMV Sem.*, pages 43–73. Birkhäuser, Basel, 2000.

[53] Richard Z. Goldstein and Edward C. Turner. A formula for Stiefel-Whitney homology classes. *Proc. Amer. Math. Soc.*, 58:339–342, 1976.

[54] Henry W. Gould. Tables of Combinatorial Identities, edited by Jocelyn Quaintance. Based on Gould's Notebooks. `http://www.math.wvu.edu/~gould`, 2010.

[55] Daniel R. Grayson and Michael E. Stillman. Macaulay2, a software system for research in algebraic geometry, Version 1.3.1. `http://www.math.uiuc.edu/Macaulay2/`, 2009.

[56] Branko Grünbaum. *Convex polytopes*, volume 221 of *Graduate Texts in Mathematics*. Springer-Verlag, New York, second edition, 2003. Prepared and with a preface by Volker Kaibel, Victor Klee and Günter M. Ziegler.

[57] Lucien Guillou and Alexis Marin, editors. *À la recherche de la topologie perdue*, volume 62 of *Progress in Mathematics*. Birkhäuser Boston Inc., Boston, MA, 1986. I. Du côté de chez Rohlin. II. Le côté de Casson. [I. Rokhlin's way. II. Casson's way].

[58] Wolfgang Haken. Theorie der Normalflächen. *Acta Math.*, 105:245–375, 1961.

[59] Percy J. Heawood. Map colour theorem. *Quart. J. Math.*, 24:332–338, 1890.

[60] Heinrich Heesch. *Gesammelte Abhandlungen*. Verlag Barbara Franzbecker Didaktischer Dienst, Bad Salzdetfurth, 1986. Edited and with a foreword by Hans-Günther Bigalke.

[61] Patricia Hersh and Isabella Novik. A short simplicial h-vector and the upper bound theorem. *Discrete Comput. Geom.*, 28(3):283–289, 2002.

[62] R. Hoppe. Die regelmässigen linear begrenzten Figuren jeder Anzahl der Dimensionen. *Archiv der Mathematik und Physik*, 67:269–290, 1882.

[63] John F. P. Hudson. *Piecewise linear topology*. University of Chicago Lecture Notes prepared with the assistance of J. L. Shaneson and J. Lees. W. A. Benjamin, Inc., New York-Amsterdam, 1969.

[64] John Philip Huneke. A minimum-vertex triangulation. *J. Combin. Theory Ser. B*, 24(3):258–266, 1978.

[65] Michael Joswig. Computing invariants of simplicial manifolds. `arXiv:math/0401176v1 [math.AT]`, Preprint, 2004.

[66] Mark Jungerman and Gerhard Ringel. The genus of the n-octahedron: regular cases. *J. Graph Theory*, 2(1):69–75, 1978.

[67] Mark Jungerman and Gerhard Ringel. Minimal triangulations on orientable surfaces. *Acta Math.*, 145(1-2):121–154, 1980.

[68] Gil Kalai. Rigidity and the lower bound theorem. I. *Invent. Math.*, 88(1):125–151, 1987.

[69] Robion C. Kirby and Laurent C. Siebenmann. On the triangulation of manifolds and the Hauptvermutung. *Bull. Amer. Math. Soc.*, 75:742–749, 1969.

[70] Hellmuth Kneser. Geschlossene Flächen in dreidimensionalen Mannigfaltigkeiten. *Jahresbericht der deutschen Mathematiker-Vereinigung*, 38:248–260, 1929.

[71] Eike Preuß Konrad Polthier, Samy Khadem-Al-Charieh and Ulrich Reitebuch. JavaView visualization software. http://www.javaview.de, 1999-2006.

[72] Matthias Kreck. An inverse to the Poincaré conjecture. *Arch. Math. (Basel)*, 77(1):98–106, 2001. Festschrift: Erich Lamprecht.

[73] Wolfgang Kühnel. Tight and 0-tight polyhedral embeddings of surfaces. *Invent. Math.*, 58(2):161–177, 1980.

[74] Wolfgang Kühnel. Higher dimensional analogues of Császár's torus. *Results Math.*, 9:95–106, 1986.

[75] Wolfgang Kühnel. Triangulations of manifolds with few vertices. In *Advances in differential geometry and topology*, pages 59–114. World Sci. Publ., Teaneck, NJ, 1990.

[76] Wolfgang Kühnel. Hamiltonian surfaces in polytopes. In *Intuitive geometry (Szeged, 1991)*, volume 63 of *Colloq. Math. Soc. János Bolyai*, pages 197–203. North-Holland, Amsterdam, 1994.

[77] Wolfgang Kühnel. Manifolds in the skeletons of convex polytopes, tightness, and generalized Heawood inequalities. In *Polytopes: abstract, convex and computational (Scarborough, ON, 1993)*, volume 440 of *NATO Adv. Sci. Inst. Ser. C Math. Phys. Sci.*, pages 241–247. Kluwer Acad. Publ., Dordrecht, 1994.

[78] Wolfgang Kühnel. *Tight polyhedral submanifolds and tight triangulations*, volume 1612 of *Lecture Notes in Mathematics*. Springer-Verlag, Berlin, 1995.

[79] Wolfgang Kühnel. Centrally-symmetric tight surfaces and graph embeddings. *Beiträge Algebra Geom.*, 37(2):347–354, 1996.

[80] Wolfgang Kühnel. Tight embeddings of simply connected 4-manifolds. *Doc. Math.*, 9:401–412 (electronic), 2004.

[81] Wolfgang Kühnel and Thomas F. Banchoff. The 9-vertex complex projective plane. *Math. Intelligencer*, 5(3):11–22, 1983.

[82] Wolfgang Kühnel and Gunter Lassmann. The unique 3-neighborly 4-manifold with few vertices. *J. Combin. Theory Ser. A*, 35(2):173–184, 1983.

[83] Wolfgang Kühnel and Gunter Lassmann. Permuted difference cycles and triangulated sphere bundles. *Discrete Math.*, 162(1-3):215–227, 1996.

[84] Wolfgang Kühnel and Frank H. Lutz. A census of tight triangulations. *Period. Math. Hungar.*, 39(1-3):161–183, 1999. Discrete geometry and rigidity (Budapest, 1999).

[85] Wolfgang Kühnel and Chrsitoph Schulz. Submanifolds of the cube. In *Applied geometry and discrete mathematics*, volume 4 of *DIMACS Ser. Discrete Math. Theoret. Comput. Sci.*, pages 423–432. Amer. Math. Soc., Providence, RI, 1991.

[86] Nicolaas H. Kuiper. Immersions with minimal total absolute curvature. In *Colloque Géom. Diff. Globale (Bruxelles, 1958)*, pages 75–88. Centre Belge Rech. Math., Louvain, 1959.

[87] Nicolaas H. Kuiper. Geometry in total absolute curvature theory. In *Perspectives in mathematics*, pages 377–392. Birkhäuser, Basel, 1984.

[88] Gunter Lassmann and Eric Sparla. A classification of centrally-symmetric and cyclic 12-vertex triangulations of $S^2 \times S^2$. *Discrete Math.*, 223(1-3):175–187, 2000.

[89] Frank H. Lutz. The Manifold Page. http://www.math.tu-berlin.de/diskregeom/stellar.

[90] Frank H. Lutz. *Triangulated Manifolds with Few Vertices and Vertex-Transitive Group Actions*. Shaker Verlag, Aachen, 1999. PhD Thesis, TU Berlin.

[91] Frank H. Lutz. Triangulated Manifolds with Few Vertices: Geometric 3-Manifolds. arXiv:math/0311116v1 [math.GT], Preprint, 48 pages, 1999.

[92] Frank H. Lutz. Triangulated Manifolds with Few Vertices: Combinatorial Manifolds. arXiv:math/0506372v1 [math.CO], Preprint, 37 pages, 2005.

[93] Frank H. Lutz, Thom Sulanke, and Ed Swartz. f-vectors of 3-manifolds. *Electron. J. Comb.*, 16(2):Research paper R13, 33 p., 2009.

[94] Brendan McKay. The nauty page. http://cs.anu.edu.au/people/bdm/nauty/, 1984.

[95] Peter McMullen. The maximum numbers of faces of a convex polytope. *Mathematika*, 17:179–184, 1970.

[96] Peter McMullen. The numbers of faces of simplicial polytopes. *Israel J. Math.*, 9:559–570, 1971.

[97] Peter McMullen and Egon Schulte. *Abstract regular polytopes*, volume 92 of *Encyclopedia of Mathematics and its Applications*. Cambridge University Press, Cambridge, 2002.

[98] Peter McMullen and David W. Walkup. A generalized lower bound conjecture for simplicial polytopes. *Mathematika*, 18:264–273, 1971.

[99] John Milnor. *Morse theory*. Based on lecture notes by M. Spivak and R. Wells. Annals of Mathematics Studies, No. 51. Princeton University Press, Princeton, N.J., 1963.

[100] John Milnor. *Lectures on the h-cobordism theorem*. Notes by L. Siebenmann and J. Sondow. Princeton University Press, Princeton, N.J., 1965.

[101] John Milnor. On the relationship between the Betti numbers of a hypersurface and an integral of its Gaussian curvature (1950). In *Collected papers. Vol. 1, Geometry*, pages 15 – 26. Publish or Perish Inc., Houston, TX, 1994.

[102] August Möbius. Gesammelte Werke, Vol. 2. Verlag Hirzel, Leipzig, 1886.

[103] Edwin E. Moise. *Geometric topology in dimensions 2 and 3*. Springer-Verlag, New York, 1977. Graduate Texts in Mathematics, Vol. 47.

[104] Marston Morse. The existence of polar non-degenerate functions on differentiable manifolds. *Ann. of Math. (2)*, 71:352–383, 1960.

[105] James R. Munkres. *Elements of algebraic topology*. Addison-Wesley Publishing Company, Menlo Park, CA, 1984.

[106] Isabella Novik. Upper bound theorems for homology manifolds. *Israel J. Math.*, 108:45–82, 1998.

[107] Isabella Novik. On face numbers of manifolds with symmetry. *Adv. Math.*, 192(1):183–208, 2005.

[108] Isabella Novik and Ed Swartz. Socles of Buchsbaum modules, complexes and posets. *Advances in Mathematics*, 222(6):2059–2084, 2009.

[109] Fritz H. Obermeyer. Jenn, a visualization software for Coxeter polytopes. http://www.jenn3d.org, 2006.

[110] Erich Ossa. *Topology. A visual introduction to geometric and algebraic foundations. (Topologie. Eine anschauliche Einführung in die geometrischen und algebraischen Grundlagen.) 2nd revised ed.*, volume 42 of *Vieweg Studium: Aufbaukurs Mathematik [Vieweg Studies: Mathematics Course]*. Friedr. Vieweg & Sohn, Braunschweig, second edition, 2009.

[111] Udo Pachner. Konstruktionsmethoden und das kombinatorische Homöomorphieproblem für Triangulierungen kompakter semilinearer Mannigfaltigkeiten. *Abh. Math. Sem. Uni. Hamburg*, 57:69–86, 1987.

[112] Gerhard Ringel. Über drei kombinatorische Probleme am n-dimensionalen Würfel und Würfelgitter. *Abh. Math. Sem. Univ. Hamburg*, 20:10–19, 1955.

[113] Gerhard Ringel. Wie man die geschlossenen nichtorientierbaren Flächen in möglichst wenig Dreiecke zerlegen kann. *Math. Ann.*, 130:317–326, 1955.

[114] Gerhard Ringel. *Map color theorem*. Springer-Verlag, New York, 1974. Die Grundlehren der mathematischen Wissenschaften, Band 209.

[115] Colin P. Rourke and Brian J. Sanderson. *Introduction to piecewise-linear topology*. Springer-Verlag, New York, 1972. Ergebnisse der Mathematik und ihrer Grenzgebiete, Band 69.

[116] Nikolai Saveliev. *Lectures on the Topology of 3-Manifolds: An Introduction to the Casson Invariant*. de Gruyter Textbook, 1999.

[117] Christoph Schulz. *Mannigfaltigkeiten mit Zellzerlegung im Randkomplex eines konvexen Polytops und verallgemeinerte Hamilton-Kreise*. Dissertation, Universität Bochum, 1974.

[118] Christoph Schulz. Polyhedral manifolds on polytopes. *Rend. Circ. Mat. Palermo (2) Suppl.*, (35):291–298, 1994. First International Conference on Stochastic Geometry, Convex Bodies and Empirical Measures (Palermo, 1993).

[119] Herbert Seifert and William Threlfall. *Seifert and Threlfall: a textbook of topology*, volume 89 of *Pure and Applied Mathematics*. Academic Press Inc. [Harcourt Brace Jovanovich Publishers], New York, 1980. Translated from the German edition of 1934 by Michael A. Goldman, With a preface by Joan S. Birman, With "Topology of 3-dimensional fibered spaces" by Seifert, Translated from the German by Wolfgang Heil.

[120] Carlos H. Séquin. Symmetrical Hamiltonian Manifolds on Regular 3D and 4D Polytopes. The Coxeter Day, Banff, Canada, Aug. 3, 2005, pp. 463–472, see http://www.cs.berkeley.edu/~sequin/BIO/pubs.html, 2005.

[121] Leonard H. Soicher. GRAPE - GRaph Algorithms using PErmutation groups, a GAP package, Version 4.3. `http://www.gap-system.org/Packages/grape.html`, 2006.

[122] Duncan M'Laren Young Sommerville. The relations connecting the angle-sums and volume of a polytope in space of n dimensions. *Proc. Royal Society London, Ser. A*, 115:103–119, 1927.

[123] Edwin H. Spanier. *Algebraic topology*. Springer-Verlag, New York, 1981. Corrected reprint.

[124] Eric Sparla. *Geometrische und kombinatorische Eigenschaften triangulierter Mannigfaltigkeiten*. Berichte aus der Mathematik. [Reports from Mathematics]. Verlag Shaker, Aachen, 1997. Dissertation, Universität Stuttgart.

[125] Eric Sparla. An upper and a lower bound theorem for combinatorial 4-manifolds. *Discrete Comput. Geom.*, 19(4):575–593, 1998.

[126] Eric Sparla. A new lower bound theorem for combinatorial $2k$-manifolds. *Graphs Combin.*, 15(1):109–125, 1999.

[127] Jonathan Spreer. Normal surfaces as combinatorial slicings. `arXiv:1004.0872v1 [math.CO]`, preprint, 28 pages, 12 figures, 2010.

[128] Jonathan Spreer. Surfaces in the cross polytope. `arXiv:1009.2642v1 [math.CO]`, preprint, 12 pages, 1 figure, 2010.

[129] Richard P. Stanley. The upper bound conjecture and Cohen-Macaulay rings. *Studies in Appl. Math.*, 54(2):135–142, 1975.

[130] Richard P. Stanley. The number of faces of a simplicial convex polytope. *Adv. in Math.*, 35(3):236–238, 1980.

[131] Richard P. Stanley. The number of faces of simplicial polytopes and spheres. In *Discrete geometry and convexity (New York, 1982)*, volume 440 of *Ann. New York Acad. Sci.*, pages 212–223. New York Acad. Sci., New York, 1985.

[132] Richard P. Stanley. Generalized *h*-vectors, intersection cohomology of toric varieties, and related results. In *Commutative algebra and combinatorics (Kyoto, 1985)*, volume 11 of *Adv. Stud. Pure Math.*, pages 187–213. North-Holland, Amsterdam, 1987.

[133] Richard P. Stanley. *Enumerative combinatorics. Vol. 1*, volume 49 of *Cambridge Studies in Advanced Mathematics*. Cambridge University Press, Cambridge, 1997. With a foreword by Gian-Carlo Rota, Corrected reprint of the 1986 original.

[134] Norman E. Steenrod. The classification of sphere bundles. *Ann. of Math. (2)*, 45:294–311, 1944.

[135] Ed Swartz. Face enumeration - from spheres to manifolds. *J. Europ. Math. Soc.*, 11:449–485, 2009.

[136] Gudlaugur Thorbergsson. Tight immersions of highly connected manifolds. *Comment. Math. Helv.*, 61(1):102–121, 1986.

[137] David W. Walkup. The lower bound conjecture for 3- and 4-manifolds. *Acta Math.*, 125:75–107, 1970.

[138] George W. Whitehead. *Elements of homotopy theory*, volume 61 of *Graduate Texts in Mathematics*. Springer-Verlag, New York, 1978.

[139] Günter M. Ziegler. *Lectures on polytopes*, volume 152 of *Graduate Texts in Mathematics*. Springer-Verlag, New York, 1995.

List of Figures

List of Tables

CURRICULUM VITAE

FELIX CHRISTIAN EFFENBERGER
geboren am 1. Mai 1983 in Frankfurt am Main

Schulbildung und Studium

1989 – 1993	Besuch der Grundschule Stuttgart Birkach.
1993 – 2002	Besuch des Wilhelms-Gymnasiums Stuttgart-Degerloch.
7/2002	Abitur am Wilhelms-Gymnasium. Leistungsfächer: Mathematik, Englisch.
2002 – 2007	Studium der Mathematik und Informatik an der Universität Stuttgart.
7/2007	Diplom in Mathematik, "mit Auszeichnung bestanden". Titel der Diplomarbeit: "Topology-based vector field visualization on 2-manifolds". Betreuer: Professor Daniel Weiskopf (SFU), Prof. Dr. Wolfgang Kühnel (Stuttgart).
2007 – 2010	Promotionsstudium im Fach Mathematik an der Universität Stuttgart. Schwerpunkt: kombinatorische Topologie. Doktorvater: Prof. Dr. Wolfgang Kühnel.
2008 – 2010	DFG Projektstelle am DFG Projekt Ku 1203/5-2.

Auslandsaufenthalte

5-12/2006	Simon Fraser University (SFU), Burnaby, BC, Canada, bei Professor Daniel Weiskopf.
3/2010	Cornell University, Ithaca, NY, USA, bei Professor Edward Swartz.

Stipendien und Preise

2001	Endrundenteilnehmer des 19. Bundeswettbewerbs Informatik, Sonderpreis der Dr. Steinfels Sprachreisen GmbH.
2004 – 2007	Stipendiat der Studienstiftung des deutschen Volkes.
5-12/2006	Auslandsstipendium der Studienstiftung.
2008 – 2010	Promotionsstipendiat der Studienstiftung.
3/2010	DAAD Kurzstipendium.
7/2010	"Best Software Presentation Award" (zusammen mit Jonathan Spreer) der Fachgruppe Computeralgebra bei der ISSAC 2010 in München.